Applications of the Mathematical Sciences in Molecular Biology

Eric S. Lander and
Michael S. Waterman, Editors

Committee on the Mathematical Sciences in
Genome and Protein Structure Research

Board on Mathematical Sciences

Commission on Physical Sciences, Mathematics, and Applications

National Research Council

NATIONAL ACADEMY PRESS
Washington, D.C. 1995

National Academy Press • 2101 Constitution Avenue, N.W. • Washington, D.C. 20418

NOTICE: The project that is the subject of this report was approved by the Governing Board of the National Research Council, whose members are drawn from the councils of the National Academy of Sciences, the National Academy of Engineering, and the Institute of Medicine. The members of the committee responsible for the report were chosen for their special competences and with regard for appropriate balance.

This report has been reviewed by a group other than the authors according to procedures approved by a Report Review Committee consisting of members of the National Academy of Sciences, the National Academy of Engineering, and the Institute of Medicine.

The National Research Council established the Board on Mathematical Sciences in 1984. The objectives of the Board are to maintain awareness and active concern for the health of the mathematical sciences and to serve as the focal point in the National Research Council for issues connected with the mathematical sciences. The Board holds symposia and workshops and prepares reports on emerging issues and areas of research and education, conducts studies for federal agencies, and maintains liaison with the mathematical sciences communities, academia, professional societies, and industry.

Support for this project was provided by the Fondation des Treilles, Alfred P. Sloan Foundation, National Science Foundation, Department of Energy, and National Library of Medicine.

Library of Congress Cataloging-in-Publication Data

Calculating the secrets of life : applications of the mathematical
sciences in molecular biology / Eric S. Lander, editor.
p. cm.
Includes bibliographical references and index.
ISBN 0-309-04886-9
1. Genetics -- Mathematical models. 2. Genetics -- Statistical methods. 3. Molecular biology -- Mathematical models. 4. Molecular biology -- Statistical methods. I. Lander, Eric S.
QH438.4.M3C35 1994
574.8'8'0151--dc20 94-37628
CIP

Printed in the United States of America.

COMMITTEE ON THE MATHEMATICAL SCIENCES IN GENOME AND PROTEIN STRUCTURE RESEARCH

ERIC S. LANDER, Whitehead Institute for Biomedical Research and Massachusetts Institute of Technology, *Chair*
WALTER GILBERT, Harvard University
HERBERT HAUPTMAN, Medical Foundation of Buffalo
MICHAEL S. WATERMAN, University of Southern California
JAMES H. WHITE, University of California at Los Angeles

Staff
JOHN R. TUCKER, Director
RUTH E. O'BRIEN, Staff Associate

BOARD ON MATHEMATICAL SCIENCES

COMMISSION ON PHYSICAL SCIENCES, MATHEMATICS, AND APPLICATIONS

RICHARD N. ZARE, Stanford University, *Chair*
RICHARD S. NICHOLSON, American Association for the Advancement of Science, *Vice-Chair*
STEPHEN L. ADLER, Institute for Advanced Study
SYLVIA T. CEYER, Massachusetts Institute of Technology
SUSAN L. GRAHAM, University of California at Berkeley
ROBERT J. HERMANN, United Technologies Corporation
RHONDA J. HUGHES, Bryn Mawr College
SHIRLEY A. JACKSON, Rutgers University
KENNETH I. KELLERMANN, National Radio Astronomy Observatory
HANS MARK, University of Texas at Austin
THOMAS A. PRINCE, California Institute of Technology
JEROME SACKS, National Institute of Statistical Sciences
L.E. SCRIVEN, University of Minnesota
A. RICHARD SEEBASS III, University of Colorado at Boulder
LEON T. SILVER, California Institute of Technology
CHARLES P. SLICHTER, University of Illinois at Urbana-Champaign
ALVIN W. TRIVELPIECE, Oak Ridge National Laboratory
SHMUEL WINOGRAD, IBM T.J. Watson Research Center
CHARLES A. ZRAKET, MITRE Corporation (retired)

NORMAN METZGER, Executive Director

The National Academy of Sciences is a private, nonprofit, self-perpetuating society of distinguished scholars engaged in scientific and engineering research, dedicated to the furtherance of science and technology and to their use for the general welfare. Upon the authority of the charter granted to it by the Congress in 1863, the Academy has a mandate that requires it to advise the federal government on scientific and technical matters. Dr. Bruce Alberts is president of the National Academy of Sciences.

The National Academy of Engineering was established in 1964, under the charter of the National Academy of Sciences, as a parallel organization of outstanding engineers. It is autonomous in its administration and in the selection of its members, sharing with the National Academy of Sciences the responsibility for advising the federal government. The National Academy of Engineering also sponsors engineering programs aimed at meeting national needs, encourages education and research, and recognizes the superior achievements of engineers. Dr. Robert M. White is president of the National Academy of Engineering.

The Institute of Medicine was established in 1970 by the National Academy of Sciences to secure the services of eminent members of appropriate professions in the examination of policy matters pertaining to the health of the public. The Institute acts under the responsibility given to the National Academy of Sciences by its congressional charter to be an advisor to the federal government and, upon its own initiative, to identify issues of medical care, research, and education. Dr. Kenneth I. Shine is president of the Institute of Medicine.

The National Research Council was organized by the National Academy of Sciences in 1916 to associate the broad community of science and technology with the Academy's purposes of furthering knowledge and advising the federal government. Functioning in accordance with general policies determined by the Academy, the Council has become the principal operating agency of both the National Academy of Sciences and the National Academy of Engineering in providing services to the government, the public, and the scientific and engineering communities. The Council is administered jointly by both Academies and the Institute of Medicine. Dr. Bruce Alberts and Dr. Robert M. White are chairman and vice chairman, respectively, of the National Research Council.

Preface

Molecular biology represents one of the greatest intellectual syntheses in the twentieth century. It has fused the traditional disciplines of genetics and biochemistry into an agent for understanding virtually any problem in biology or medicine. Moreover, it has produced a set of powerful techniques—called recombinant DNA technology—applicable to fundamental research and to biological engineering.

Even as molecular biology establishes itself as the dominant paradigm throughout biology, the field itself is undergoing a new and profound transformation. With the availability of ever more powerful tools, molecular biologists have begun to assemble massive databases of information about the structure and function of genes and proteins. It is becoming clear that it will soon be possible to catalogue virtually all genes and to identify virtually all basic protein structures. What began as an enterprise akin to butterfly collecting has become an effort to construct biology's equivalent of the Periodic Table: a complete delineation of the molecular building blocks of life on this planet. The new thrust is most obvious in the Human Genome Project,[1] but it is paralleled by similarly oriented efforts in structural and functional biology as well.

As molecular biology works toward characterizing the genetic basis of biological processes, mathematical and computational sciences are beginning to play an increasingly important role: they will be essential for organization, interpretation, and prediction of the burgeoning experimental information. The role of mathematical theory in biology is, to be sure, different from its role in physics (which is more amenable to description by a set of simple equations), but it is no less crucial.

The National Research Council organized the Committee on the Mathematical Sciences in Genome and Protein Structure Research to evaluate whether there was a need for increased interaction between mathematics and molecular biology. In its initial meeting, the committee

---

[1]Dausset, J., and H. Cann, 1994, "Our Genetic Patrimony," *Science* **264** (September 30), 1991; National Research Council, 1988, *Mapping and Sequencing the Human Genome*, Washington, D.C.: National Academy Press.

unanimously agreed that a need was evident. Focusing on the impediments to progress in the area, the committee concluded that the greatest obstacle to progress at the interface of these fields was not a lack of talented mathematicians, talented biologists, or grant funding. Rather, the major barrier was communication: mathematicians interested in working on problems in molecular biology faced an uphill battle in learning about a completely new and fast-moving field. In most cases, researchers working successfully at the interface of mathematics and molecular biology had solved this problem by finding a colleague willing to invest considerable time to teach them enough to be able to identify important problems and to begin productive work.

The committee decided that it could make its greatest contribution not by writing a report confirming the need for interactions between mathematics and molecular biology, but rather by (to put it in biological terms) lowering the activation energy barrier for those interested in working at the interface. Specifically, the committee members agreed to produce a book that could serve as an introduction to the interface between mathematics and molecular biology.

This book of signed chapters is the result of some three years of effort to create a product that would be interesting and accessible to both mathematicians and biologists. The book is not intended as a textbook, but rather as an introduction and an invitation to learn more. Each chapter aims to describe an important biological problem to which mathematical methods have made a significant contribution. As the examples make clear, mathematical and statistical issues have contributed key insights and advances to molecular biology, and, conversely, molecular biology has posed new challenges in the mathematical sciences. The book highlights those areas of the mathematical, statistical, and computational sciences that are important in cutting-edge research in molecular biology. It also tries to illustrate to the molecular biology community the role of mathematical methodologies in solving biomolecular problems.

Although there is a growing community of researchers working at the interface of molecular biology and the mathematical sciences, the need still far outstrips the supply. The Board on Mathematical Sciences hopes this book will inspire more individuals to become involved.

This book would not have been possible without sustained efforts by a number of people, to whom the committee and the Board on

Mathematical Sciences are grateful: John Tucker, Lawrence Cox, Hans Oser, and John Lavery played key roles in coordinating the study. Ruth O'Brien, Roseanne Price, and Susan Maurizi edited the text and oversaw production. Anonymous reviewers contributed to the clarity and understanding of the final text. The Alfred P. Sloan Foundation, the National Science Foundation, the Department of Energy, and the National Library of Medicine provided financial support. The Fondation des Treilles hosted and supported a week-long meeting at which the committee members presented extended lectures that became the basis for most of the chapters here. The committee wishes to thank all of these people and organizations for their assistance.

Contents

Applications of the Mathematical Sciences in Molecular Biology

Chapter 1
The Secrets of Life: A Mathematician's Introduction to Molecular Biology

Eric S. Lander
Whitehead Institute for Biomedical Research
and Massachusetts Institute of Technology

Michael S. Waterman
University of Southern California

Molecular biology has emerged from the synthesis of two complementary approaches to the study of life—biochemistry and genetics—to become one of the most exciting and vibrant scientific fields at the end of the twentieth century. This introductory chapter provides a brief history of the intellectual foundations of modern molecular biology and defines key terms and concepts that recur throughout the subsequent chapters.

The concepts of molecular biology have become household words. DNA, RNA, and enzymes are routinely discussed in newspaper stories, prime-time television shows, and business weeklies. The passage into popular culture is complete only 40 years after the discovery of the structure of deoxyribonucleic acid (DNA) by James Watson and Francis Crick and only 20 years after the first steps toward genetic engineering. With breathtaking speed, these basic scientific discoveries have led to astonishing scientific and practical implications: the fundamental biochemical processes of life have been laid bare. The evolutionary record of life can be read from DNA sequences. Genes for proteins such as human insulin can be inserted into bacteria, which then can inexpensively produce large and pure amounts of the protein. Farm animals and crops can be engineered to produce healthier and more desirable products. Sensitive and reliable diagnostics can be developed for viral diseases such as AIDS, and treatments can be developed for some hereditary diseases, such as cystic fibrosis.

Molecular biology is certain to continue its exciting growth well into the next century. As its frontiers expand, the character of the field is changing. With ever growing databases of DNA and protein sequences and increasingly powerful techniques for investigating structure and function, molecular biology is becoming not just an experimental science, but a *theoretical* science as well. The role of theory in molecular biology is not likely to resemble the role of theory in physics, in which mathematicians can offer grand unifying theories. In biology, key insights emerge less often from first principles than from interpreting the crazy quilt of solutions that evolution has devised. Interpretation depends on having theoretical tools and frameworks. Sometimes, these constructs are nonmathematical. Increasingly, however, the mathematical sciences—mathematics, statistics, and computational science—are playing an important role.

This book emerged from the recognition of the need to cultivate the interface between molecular biology and the mathematical sciences. In the following chapters, various mathematicians working in molecular biology provide glimpses of that interface. The essays are not intended to be comprehensive up-to-date reviews, but rather vignettes that describe just enough to tempt the reader to learn more about fertile areas for research in molecular biology.

This introductory chapter briefly outlines the intellectual foundations of molecular biology, introduces some key terms and concepts that recur throughout the book, and previews the chapters to follow.

BIOCHEMISTRY

Historically, molecular biology grew out of two complementary experimental approaches to studying biological function: biochemistry and genetics (Figure 1.1). Biochemistry involves fractionating (breaking up) the molecules in a living organism, with the goal of purifying and characterizing the chemical components responsible for carrying out a particular function. To do this, a biochemist devises an assay for

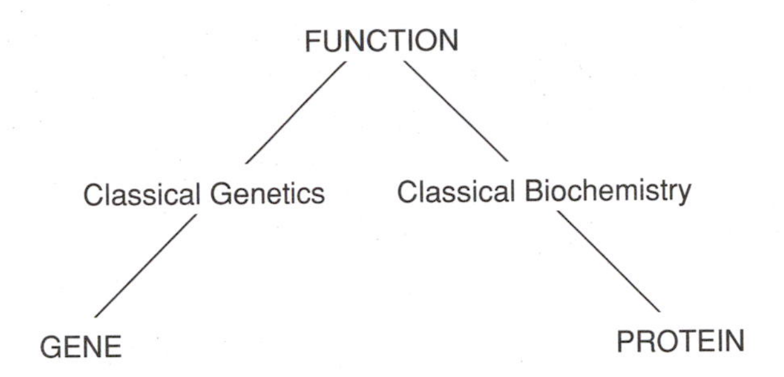

FIGURE 1.1 Genetics and biochemistry began as independent ways to study biological function.

measuring an "activity" and then tries successive fractionation procedures to isolate a pure fraction having the activity. For example, a biochemist might study an organism's ability to metabolize sugar by purifying a component that could break down sugar in a test tube.

In vitro (literally, in glass) assays were accomplished back in the days when biologists were still grappling with the notion of vitalism. Originally, it was thought that life and biochemical reactions did not obey the known laws of chemistry and physics. Such vitalism held sway until about 1900, when it was shown that material from dead yeast cells could ferment sugar into ethanol, proving that important processes of living organisms were "just chemistry." The catalysts promoting these transformations were called enzymes.

Living organisms are composed principally of carbon, hydrogen, oxygen, and nitrogen; they also contain small amounts of other key elements (such as sodium, potassium, magnesium, sulfur, manganese, and selenium). These elements are combined in a vast array of complex macromolecules that can be classified into a number of major types: proteins, nucleic acids, lipids (fats), and carbohydrates (starches and sugars). Of all the macromolecules, the proteins have the most diverse range of functions. The human body makes about 100,000 distinct proteins, including:

- enzymes, which catalyze chemical reactions, such as digestion of food;
- structural molecules, which make up hair, skin, and cell walls;

- transporters of substances, such as hemoglobin, which carries oxygen in blood; and
- transporters of information, such as receptors in the surface of cells and insulin and other hormones.

In short, proteins do the work of the cell. From a structural standpoint, a protein is an ordered linear chain made of building blocks known as amino acids (Figures 1.2 and 1.3). There are 20 distinct amino acids, each with its own chemical properties (including size, charge, polarity, and hydrophobicity, or the tendency to avoid packing with water). Each protein is defined by its unique sequence of amino acids; there are typically 50 to 500 amino acids in a protein.

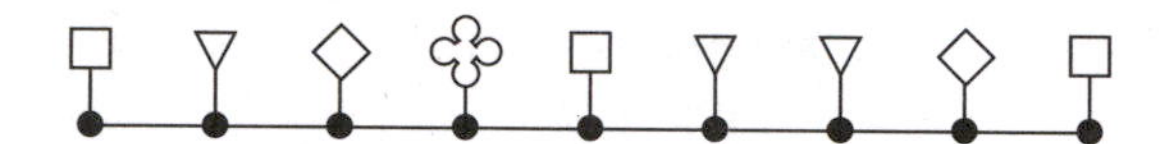

FIGURE 1.2 Proteins are a linear polymer, assembled from 20 building blocks called amino acids that differ in their side chains. The diagram shows a highly stylized view of this linear structure.

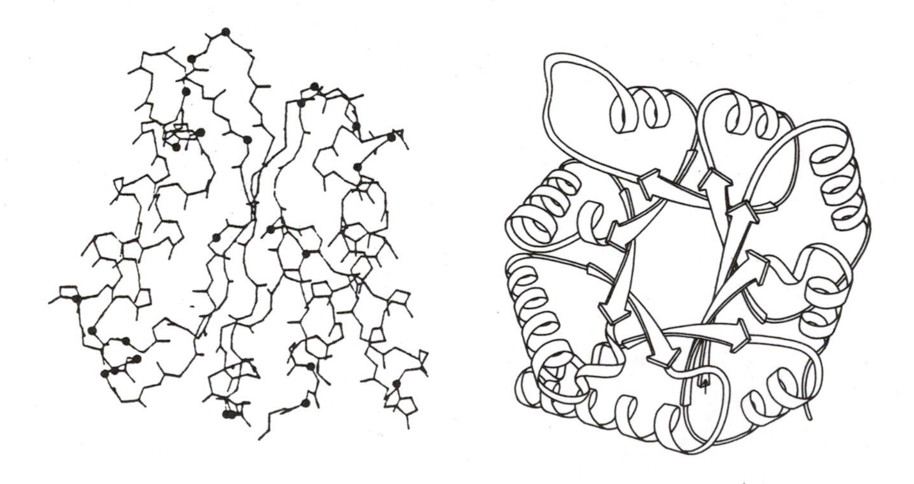

FIGURE 1.3 Examples of different representations of protein structures focusing on (left) chemical bonds and (right) secondary structural features such as helices and sheet-like elements. Reprinted, by permission, from Richardson and Richardson (1989). Copyright © 1989 by the Plenum Publishing Corporation.

The amino acid sequence of a protein causes it to fold into the particular three-dimensional shape having the lowest energy. This gives the protein its specific biochemical properties, that is, its function. Typically, the shape of a protein is quite robust. If the protein is heated, it will be denatured (that is, lose its three-dimensional structure), but it will often reassume that structure (refold) when cooled. Predicting the folded structure of a protein from the amino acid sequence remains an extremely challenging problem in mathematical optimization. The challenge is created by the combinatorial explosion of plausible shapes, each of which represents a local minimum of a complicated nonconvex function of which the global minimum is sought.

CLASSICAL GENETICS

The second major approach to studying biological function has been genetics. Whereas biochemists try to study one single component purified away from the organism, geneticists study mutant organisms that are intact except for a single component. Thus a biochemist might study an organism's ability to metabolize sugar by finding mutants that have lost the ability to grow using sugar as a food source.

Genetics can be traced back to the pioneering experiments of Gregor Mendel in 1865. These key experiments elegantly illustrate the role of theory and abstraction in biology. For his experiments, Mendel started with **pure breeding** strains of peas—that is, ones for which all offspring, generation after generation, consistently show a trait of interest. This choice was key to interpreting the data.

One of the traits that he studied was whether the pea made round or wrinkled seeds. Starting with pure breeding round and wrinkled strains, Mendel made a controlled cross to produce an F_1 generation. (The *i*th generation of the cross is denoted F_i.) Mendel noted that all of the F_1 generation consisted of round peas; the wrinkled trait had completely vanished. However, when Mendel crossed these F_1 peas back to the pure breeding wrinkled parent, the wrinkled trait reappeared: of the second generation, approximately half were round and half were wrinkled. Moreover, when Mendel crossed the F_1 peas to themselves, he found that

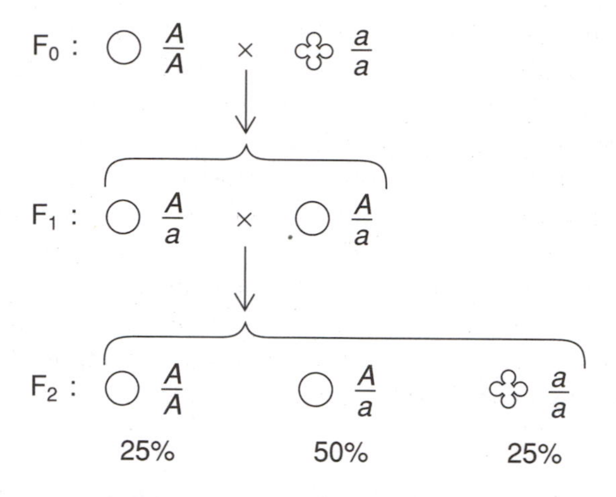

FIGURE 1.4 Mendel's crosses between pure breeding peas with round and wrinkled seeds revealed the telltale binomial ratio 1:2:1 in the second generation that led Mendel to infer the existence of discrete particles of inheritance.

the second generation showed 75 percent round and 25 percent wrinkled (Figure 1.4).

On the basis of these and other experiments, Mendel hypothesized that traits such as roundness are affected by discrete factors—which today we call genes. In particular, Mendel suggested the following:

- Each organism inherits two copies of a gene, one from each parent. Each parent passes on one of the two copies, chosen at random, to each offspring. (These important postulates are called Mendel's First Law of Inheritance.)
- Genes can occur in alternative forms, called **alleles**. For example, the gene affecting seed shape occurs in one form (allele *A*) causing roundness and one form (allele *a*) causing wrinkledness.
- The pure breeding round and wrinkled plants carried two copies of the same allele, *AA* and *aa*, respectively. Individuals carrying two copies of the same gene are called **homozygotes**. The F_1 generation consists of individuals with genotype *Aa*, with the round trait dominant over the wrinkled trait. Such individuals are called **heterozygotes**.

- In the cross of the F_1 generation (*Aa*) to the pure breeding wrinkled strain (*aa*), the offspring were a 1:1 mixture of *Aa:aa* according to which allele was inherited from the F_1 parent. In the cross between two F_1 parents (*Aa*), the offspring were a 1:2:1 mixture of *AA:Aa:aa* according to the binomial selection of alleles from the two parents.

It is striking to realize that the existence of genes was deduced in this abstract mathematical way. Probability and statistics were an intrinsic part of early genetics, and they have remained so. Of course, Mendel did not have formal statistical analysis at his disposal, but he managed to grasp the key concepts intuitively. Incidentally, the famous geneticist and statistician R.A. Fisher analyzed Mendel's data many years later and concluded that they fit statistical expectation a bit too well. Mendel probably discarded some outliers as likely experimental errors.

It was almost 35 years before biologists had an inkling of where these hypothetical genes resided in the cell (in the chromosomes) and almost 100 years before they understood their biochemical nature.

MOLECULAR BIOLOGY

As suggested in Figure 1.1, the biochemical and the genetic approaches were virtually disjoint: the biochemist primarily studied proteins, whereas the geneticist primarily studied genes. Much like the great unifications in mathematics, molecular biology emerged from the recognition that the two apparently unrelated fields were, in fact, complementary perspectives on the same subject.

The first clues emerged from the study of mutant microorganisms in which gene defects rendered them unable to synthesize certain key macromolecules. Biochemical study of these genetic mutants showed that each lacked a specific enzyme. From these experiments the hypothesis became clear that genes somehow must "encode" enzymes. This (Nobel-Prize-winning) notion was dubbed the "one gene-one enzyme" hypothesis, although today it has been modified to "one gene-one protein." Of course, the mystery remained: How do genes encode proteins?

The answer depended on finding the biochemical nature of the gene itself, thereby uniting the fields. To purify the gene as a biochemical entity, one needed a test tube assay for heredity—something that might seem impossible. Fortunately, scientific serendipity provided a solution. In a famous series of bacteriological studies, Griffith showed 50 years ago that certain properties (such as pathogenicity) could be transferred from dead bacteria to live bacteria. Avery et al. (1944) were able to successively fractionate the dead bacteria so as to purify the elusive "transforming principle," the material that could confer new heredity on bacteria. The surprising conclusion was that the gene appeared to be made of DNA.

The notion of DNA as the material of heredity came as a surprise to most biochemists. DNA was known to be a linear polymer of four building blocks called nucleotides (referred to as adenine, thymine, cytosine, and guanine, and abbreviated as `A`, `T`, `C`, and `G`) joined by a sugar-phosphate backbone. However, most knowledgeable scientists reckoned that the polymer was a boring, repetitive structural molecule that functioned as some sort of scaffold for more important components. In the days before computers, it was not apparent how a linear polymer might encode information. If DNA contained the genes, the structure of DNA became a key issue.

In their legendary work in 1953, Watson and Crick correctly inferred the structure of most DNA and, in so doing, explained the main secret of heredity. While some viruses have single-stranded DNA, the DNA of humans and of most other forms of life consists of two antiparallel chains (strands) in the form of a double helix in which the bases (nucleotides) pair up to form **base pairs** in a certain way (Figure 1.5) so that the sequence of one chain completely specifies the sequence of the other: an `A` on one chain always corresponds to a `T` on the other, and a `G` to a `C`. The sequences are complementary. The fact that the information is redundant explains the basis for the replication of living organisms: the two strands of the double helix unwind, and each serves as a template for the synthesis of a complete double helix that is passed on to a daughter cell. This process of replication is carried out by enzymes called **DNA polymerases**. Mutations are changes in the nucleotide sequence in DNA. Mutations can be induced by external

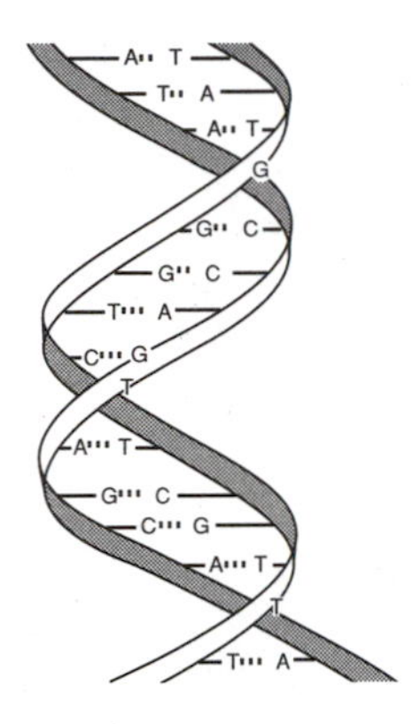

FIGURE 1.5 The DNA double helix consists of anti-parallel helical strands, with complementary bases (G–C and A–T).

forces such as sunlight and chemical agents or can occur as random copying errors during replication.

There remained the question of how the 4-letter alphabet of DNA could "encode" the instructions for the 20-letter alphabet of protein sequences. Biochemical studies over the next decade showed that genes correspond to specific stretches of DNA along a chromosome (much like individual files on a hard disk). These stretches of DNA can be expressed at particular times or under particular circumstances. Typically, gene expression begins with **transcription** of the DNA sequence into a messenger molecule made of ribonucleic acid (RNA) (Figure 1.6A). This transcription process is carried out by enzymes called **RNA polymerases**. RNA is structurally similar to DNA and consists of four building blocks, the nucleotides denoted A, U, C, and G, with U (uracil) playing the role of T. The **messenger RNA** (mRNA) is copied from the DNA of a gene according to the usual base pairing rules (a U in RNA corresponds to an A in DNA, an A corresponds to a T, a G to a C, and a C to a G). The messenger RNA copied from a gene is single-stranded and is just an unstable intermediate used for transmitting information from the cell nucleus (where the DNA resides) to the cytoplasm (where protein synthesis occurs). The mRNA is then translated into a protein by a remarkable molecular machine called the **ribosome.**

A

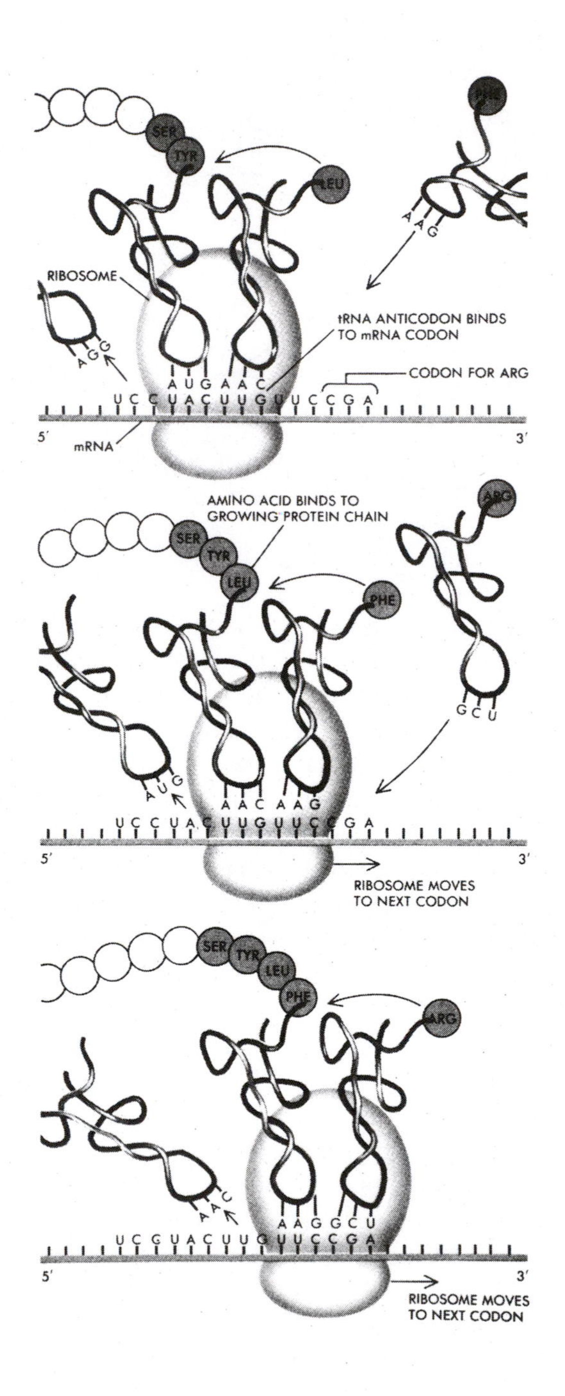
SER
TYR
LEU
PHE
RIBOSOME
tRNA ANTICODON BINDS TO mRNA CODON
CODON FOR ARG
AUGAAC
UCCUACUUGUUCCGA
5′
mRNA
3′
AMINO ACID BINDS TO GROWING PROTEIN CHAIN
ARG
GCU
AUG
AACAAG
RIBOSOME MOVES TO NEXT CODON
AAGGCU
UCGUACUUGUUCCGA
RIBOSOME MOVES TO NEXT CODON

B

FIRST POSITION (5′ END)	SECOND POSITION U	C	A	G	THIRD POSITION (3′ END)
U	Phe	Ser	Tyr	Cys	U
	Phe	Ser	Tyr	Cys	C
	Leu	Ser	Stop	Stop	A
	Leu	Ser	Stop	Trp	G
C	Leu	Pro	His	Arg	U
	Leu	Pro	His	Arg	C
	Leu	Pro	Gln	Arg	A
	Leu	Pro	Gln	Arg	G
A	Ile	Thr	Asn	Ser	U
	Ile	Thr	Asn	Ser	C
	Ile	Thr	Lys	Arg	A
	Met	Thr	Lys	Arg	G
G	Val	Ala	Asp	Gly	U
	Val	Ala	Asp	Gly	C
	Val	Ala	Glu	Gly	A
	Val	Ala	Glu	Gly	G

Note: Given the position of the bases in a codon, it is possible to find the corresponding amino acid. For example, the codon (5′) AUG (3′) on mRNA specifies methionine, whereas CAU specifies histidine. UAA, UAG, and UGA are termination signals. AUG is part of the initiation signal, and it codes for internal methionines as well.

FIGURE 1.6 After messenger RNA is transcribed from the DNA sequence of a gene, it is translated into protein by a remarkable molecular device called the ribosome. (A) Ribosomes read the RNA bases and write a corresponding amino acid sequence. The correct amino acid is brought into juxtaposition with the correct nucleotide triplet through the mediation of an adapter molecule known as transfer RNA. (B) The table showing the correspondence between triplets of bases and amino acids is called the genetic code. Reprinted from *Recombinant DNA: A Short Course* by Watson, Tooze, and Kurtz (1994). Copyright © 1994 James D. Watson, John Tooze, and David T. Kurtz. Used with permission of W.H. Freeman and Company.

The ribosome "reads" the linear sequence of the mRNA and "writes" (i.e., creates) a corresponding linear sequence of amino acids of the encoded protein. Translation is carried out according to a three-letter code: a group of three letters is a **codon** that specifies a particular amino acid according to a look-up table called the genetic code (Figure 1.6B). There are 4^3 different codons. The codons are read in contiguous, nonoverlapping fashion from a defined starting point, called the translational start site. Finally, the newly synthesized amino acid chain spontaneously folds into its three-dimensional structure. (For a recent discussion of protein folding, see Sali et al., 1994.)

The details of the genetic code were solved by elegant biochemical tricks, which were necessary because chemists had only the ability to synthesize random collections of RNA having defined proportions of different bases. With some combinatorial reasoning, this proved to be sufficient. For example, if the ribosome is given an mRNA with the sequence `UUUUU...`, then it makes a protein chain consisting of only the amino acid phenylalanine (Phe). Thus `UUU` must encode phenylalanine. By examining more complex mixtures, researchers soon worked out the entire genetic code.

Molecular biology provides the third leg of the triangle, relating genetics and biochemistry (Figure 1.7).

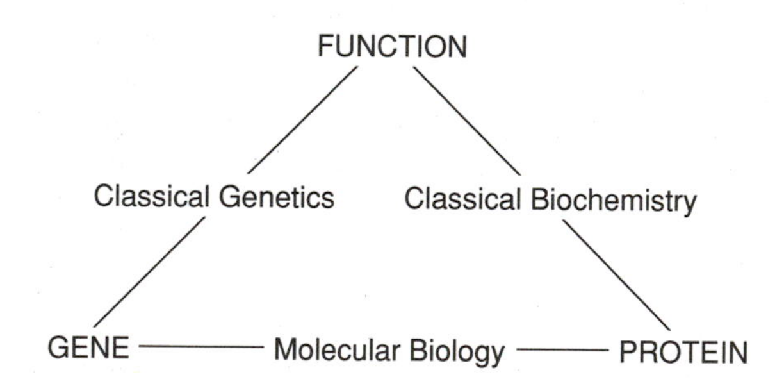

FIGURE 1.7 Molecular biology connected the disciplines of genetics and biochemistry by showing how genes encoded proteins.

THE RECOMBINANT DNA REVOLUTION

By 1965, molecular biology had laid bare the basic secrets of life. Without the ability to manipulate genes, however, the understanding was more theoretical than operational. In the 1970s, this situation was transformed by the recombinant DNA revolution.

Biochemists discovered a variety of enzymes made by bacteria that allowed one to manipulate DNA at will. Bacteria made **restriction enzymes**, which cut DNA at specific sequences and served as a defense against invading viruses, and **ligases**, which join DNA fragments. With these and other tools (which are now all readily available from commercial suppliers), it became possible to cut and paste DNA fragments at will and to introduce them into living cells (Figure 1.8). Such cloning experiments allow scientists to reproduce unlimited quantities of specific DNA molecules and have led to detailed understanding of individual genes. Moreover, producing recombinant DNA molecules that contain bacterial DNA instructions for making a particular human protein (such as insulin) gave birth to the biotechnology industry.

A key development was the invention of **DNA sequencing**, the process of determining the precise nucleotide sequence of a cloned DNA molecule. With DNA sequencing, it became possible to read the sequence of any gene in stretches of 300 to 500 nucleotides at a time. DNA sequencing has revealed striking similarities among living creatures as diverse as humans and yeast, with far-reaching consequences for our understanding of molecular structure and evolution. DNA sequencing has also led to an information explosion in biology, with public databases still expanding at a rapid exponential rate. In early 1993, there were over 100 million bases of DNA in the public databases. For reference, the entire genome of the intestinal bacteria *Escherichia coli* (*E. coli*) consists of about 4.6 million bases, and the human genome sequence has roughly 3 billion bases.

In recent years a powerful new technique called the **polymerase chain reaction** (PCR) has been added to the molecular biologist's tool kit (Figure 1.9). PCR allows one to directly amplify a specific DNA sequence without resort to cloning. To perform PCR, one uses short DNA molecules called **primers** (typically about 20 bases long) that are complementary to the sequences flanking the region of interest. Each

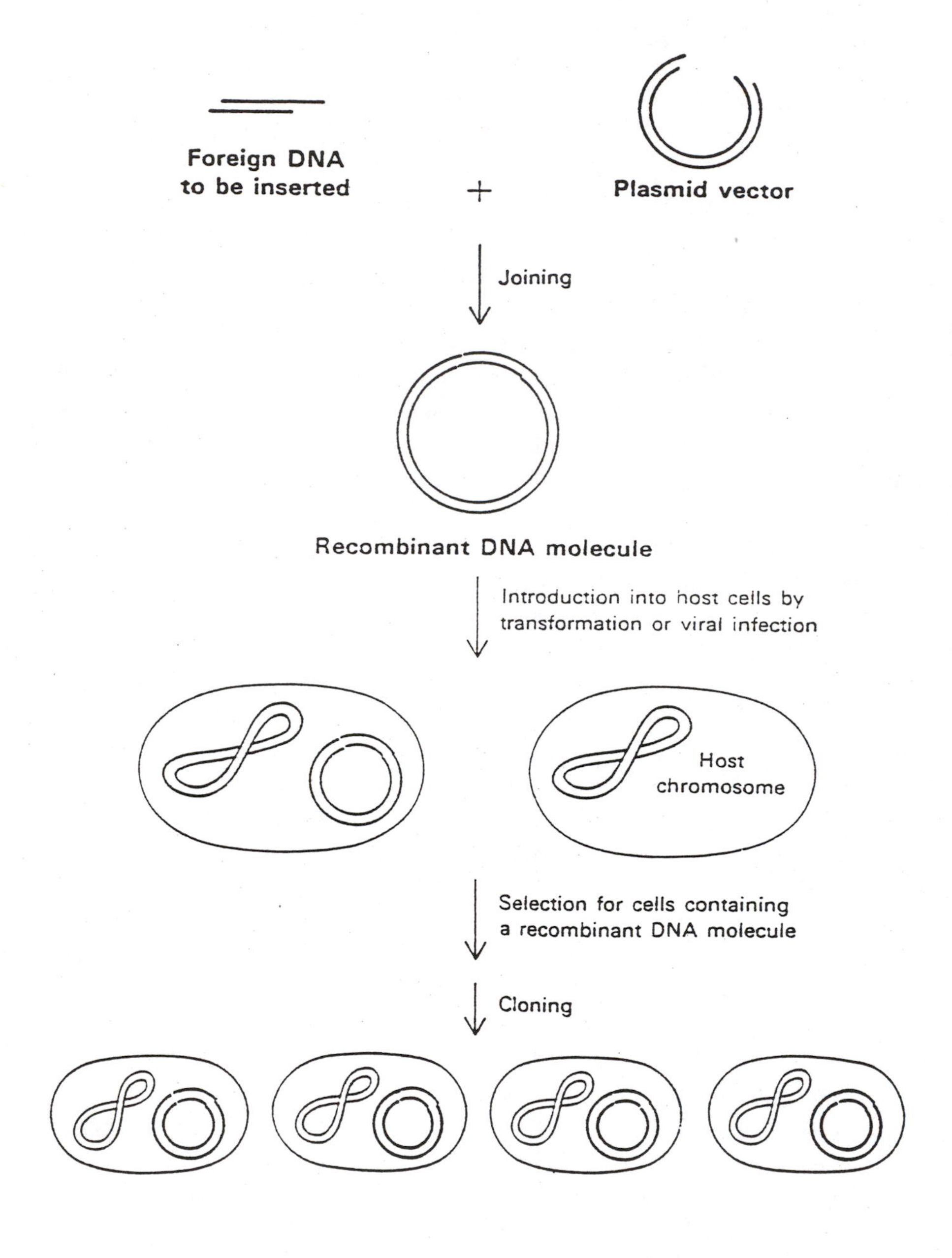

FIGURE 1.8 By cloning a foreign DNA molecule in a plasmid vector, it is possible to propagate the DNA in a bacterial or other host cell.

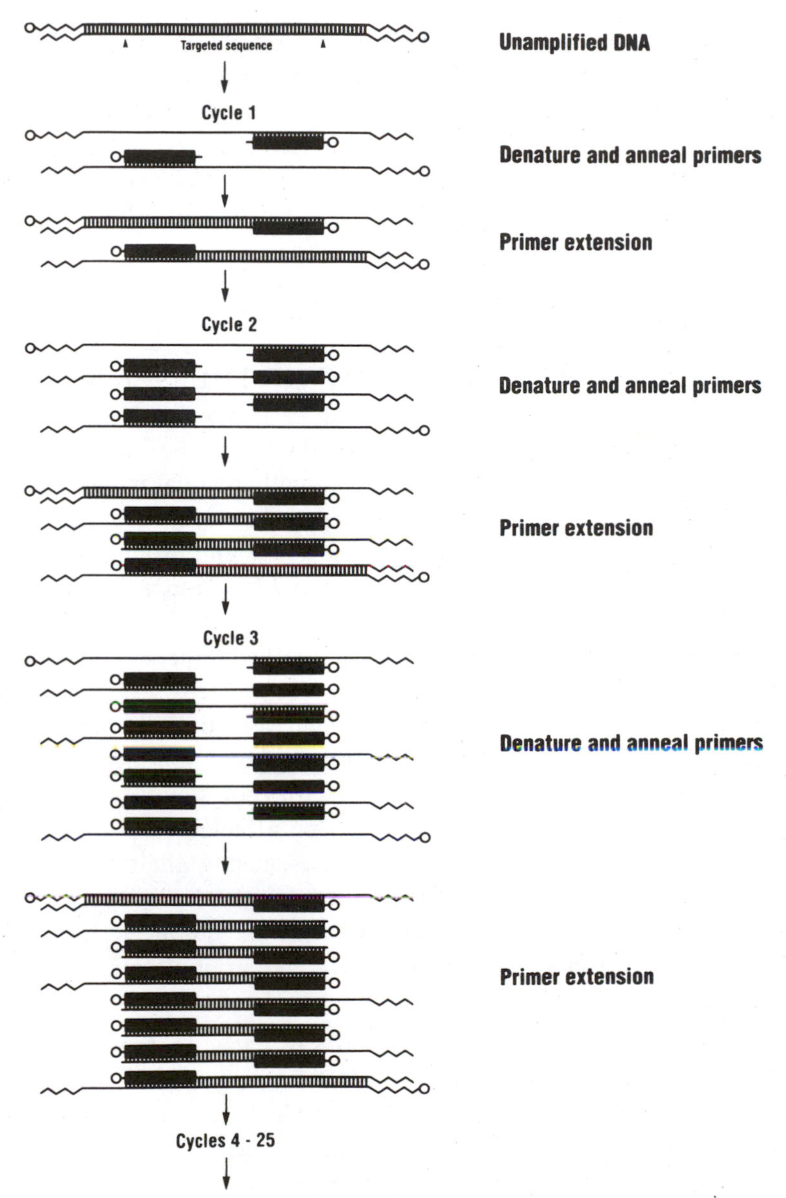

FIGURE 1.9 The polymerase chain reaction (PCR) allows exponential amplification of DNA. The method involves successive rounds of copying (using the enzyme DNA polymerase) between two synthetic primers corresponding to nearby DNA sequences. Each round doubles the number of copies. Courtesy of the Perkin-Elmer Corporation. Reprinted from the National Research Council (1992).

primer is allowed to pair with a base in the complementary region and is then extended to contain the full sequence from the region by using the enzyme DNA polymerase. In this fashion a single copy of the region gives rise to two copies. By iterating this step *n* times, one might make 2^n copies of the region. In practice, one can start with a small drop of blood or saliva and obtain a millionfold amplification of a region. Not surprisingly, PCR has found myriad applications, especially in genetic diagnostics.

MOLECULAR GENETICS IN THE 1990s

With the tools of recombinant DNA, the triangle of knowledge (see Figure 1.7) has been transformed, to use a mathematical metaphor, into a commutative diagram (Figure 1.10). It is possible to traverse the diagram in any direction—for example, to find the genes and proteins underlying a biological function or to find the protein and function associated with a given gene.

A good illustration of the power of the techniques is provided by recent studies of the inherited disease cystic fibrosis (CF). CF is a recessive disease, the genetics of which is formally identical to wrinkledness in peas as studied by Mendel: if two non-affected carriers of the recessive CF gene *a* (that is, heterozygotes with genotype *Aa*) marry, one fourth of their offspring will be affected (that is, will have genotype *aa*). The frequency of the disease-causing allele is about 1/42 in the Caucasian population, and so about 1/21 of all Caucasians are carriers. Since a marriage between two carriers produces 1/4 affected children, the disease frequency in the population is about $1/2000\ (\approx 1/4 \times 1/21 \times 1/21)$.

Although CF was recognized relatively early in the century, the molecular basis for the disease remained a mystery until 1989. The first breakthrough was the **genetic mapping** of CF to human chromosome 7 in 1985 (Figure 1.11). Genetic mapping involved showing that the inheritance pattern of the disease in families is closely correlated with the inheritance pattern of a particular **DNA polymorphism** (that is, a common spelling variation in the DNA), in this case on chromosome 7.

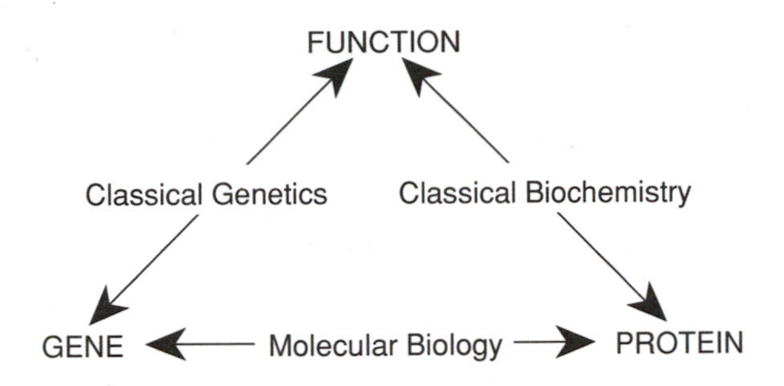

FIGURE 1.10 Recombinant DNA provided the ability to move freely in any direction among gene, protein, and function, thereby converting the triangle of Figure 1.7 into a commutative diagram.

The correlation does not imply that the polymorphism causes the disease, but rather that the polymorphism must be located near the site of the disease gene. Of course, "near" is a relative term. In this case, "near" meant that the CF gene must be within 1 million to 2 million bases of DNA along the chromosome. The next step was the physical mapping and the DNA sequencing of the CF gene itself, which took four more years to accomplish. This involved starting from the nearby polymorphism and sequentially isolating adjacent fragments in a tedious process called **chromosomal walking** until the disease gene was reached. Once the disease gene was found, its complete DNA sequence was determined. (A description of how one knows that one has found the disease gene is beyond the scope of this introduction.)

From the DNA sequence, it became clear that the CF gene encoded a protein of 1,480 amino acids and that the most common misspelling in the population (accounting for about 70 percent of all CF alleles) was a three-letter deletion that removed a single codon specifying an amino acid, a phenylalanine at position 508 of the protein. On the basis of this finding, it became possible to perform DNA diagnostics on individuals to see if they carried the common CF mutation.

Even more intriguingly, the sequence gave immediate clues to the structure and function of the gene product. When the protein sequence was compared with the public databases of previous sequences, it was found to show strong similarities to a class of proteins that were membrane-bound transporters—molecules that reside in the cell membrane, bind adenosinetriphosphate (ATP), and transport substances

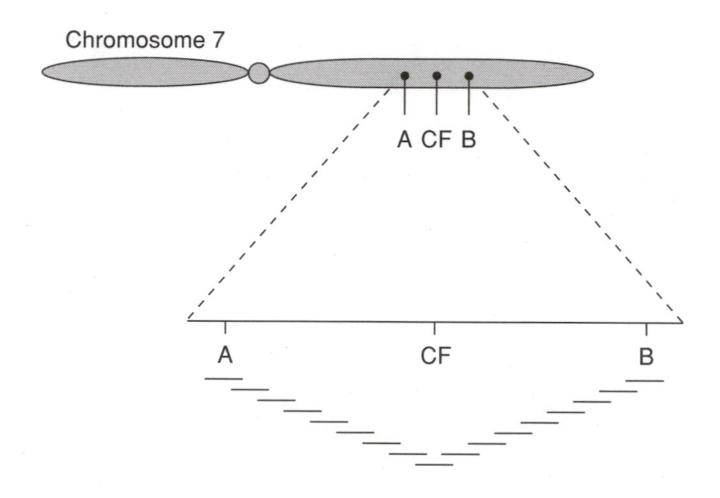

FIGURE 1.11 Chromosomal walking from flanking genetic markers to the gene responsible for cystic fibrosis. The distance covered totaled more than 1 million DNA bases.

into and out of the cell (Figure 1.12A). By analogy, it was even possible to infer a likely three-dimensional shape for the CF protein (Figure 1.12B). In this way, computer-based sequence analysis shed substantial light on the structure and function of this important disease gene.

With the recent advent of gene therapy—the ability to use a virus as a shuttle to deliver a working copy of a gene into cells carrying a defective version—clinical trials have been started to try to cure the disease in the lung cells of CF patients. The path from the initial discovery of the gene to potential therapies has been stunningly short in this case.

THE HUMAN GENOME PROJECT

With the identification of the CF gene as well as a number of other successes, it has become clear that molecular genetics has developed a powerful general paradigm that can be applied to many inherited diseases and will have a profound impact on our understanding of human health. Unfortunately, the paradigm involves many tedious laboratory steps: genetic mapping (finding a polymorphism closely linked to the disease gene), physical mapping (isolating the consecutive fragments of DNA along the chromosome), and DNA sequencing (typically performed in pieces of only 300 to 500 letters at a time). It would be inefficient to repeat these steps for each of the more than 4,000 genetic traits and diseases already known. To accelerate progress, molecular geneticists have seen the value of building infrastructure—a common set of maps, tools, and information—that can be applied to all genetic problems. This recognition led to the creation of the Human Genome Project (National Research Council, 1988), an international effort to analyze the structure of the human genome (as well as the genomes of certain key experimental model systems, such as *E. coli*, yeast, nematodes, fruit flies, and mice).

Because most molecular biological methods are applicable only to small fragments of DNA, it is not practical to sequence the human genome by simply starting at one end and proceeding sequentially. Moreover, because the current cost of sequencing is about $1 per base, it would be expensive to sequence the 3×10^9 bases of the human chromosomes by conventional methods. Instead, it is more sensible to construct maps of increasing resolution and to develop more efficient sequencing technology. The current goals of the Human Genome Project include development of the following tools:

- *Genetic maps*. The goal is to produce a genetic map showing the location of 5,000 polymorphisms that can be used to trace inheritance of diseases in families. As of this writing, the goal is nearly complete.
- *Physical maps*. The goal is to produce a collection of overlapping pieces of DNA that cover all the human chromosomes. This goal is not completed yet but should be by 1996.
- *DNA sequence*. The ultimate goal is to sequence the entire genome, but the intermediate steps include sequencing particular regions, generating more efficient and automated technology, and developing better analytical methods for handling DNA information.

With the vast quantities of information being generated, the Human Genome Project is one of the driving forces behind the expanding role

A

```
              ■ ■■■■ ▬▬▬ ▬▬▬▬▬▬▬▬▬                ■  ■  ▬▬▬▬▬*▬          ▬▬▬▬ ▬▬▬  ▬▬▬▬▬▬▬▬▬
CFTR  (N)  FSLLGTPVLKDINFKIERGQLLAVAGSTGAGKTSLLMMIMG   ISFCSQFSWIMPGTIK-ENIIFGVSYD   GEGGITLSGGQRARISLARAVYKDADLYLLDSPFGYLDVLTEK
CFTR  (C)  YTEGGNAILENISFSISPGQRVGLLGRTGSGKSTLLSAFLR   DSITLQQWRKAFGVIPQKVFIFSGTFR   VDGGCVLSHGHKQLMCLARSVLSKAKILLLDEPSAHLDPVTYQ
hmdr1 (N)  PSRKEVKILKGLNLKVQSGQTVALVGNSGCGKSTTVQLMQR   IGVVSQEPVLFATTI-AENIRYGRENV   GERGAQLSGGQKQRIAIARALVRNPKILLLDEATSALDTESEA
hmdr1 (C)  PTRPDIPVLQGLSLEVKKGQTLALVGSSGCGKSTVVQLLER   LGIVSQEPILFDCSI-AENIAYGDNSR   GDKGTLLSGGQKQRIAIARALVRQPHILLLDEATSALDTESEK
mmdr1 (N)  PSRSEVQILKGLNLKVKSGQTVALVGNSGCGKSTTVQLMQR   IGVVSQEPVLFATTI-AENIRYGREDV   GERGAQLSGGQKQRIAIARALVRNPKILLLDEATSALDTESEA
mmdr1 (C)  PTRPNIPVLQGLSLEVKKGQTLALVGSSGCGKSTVVQLLER   LGEVSQEPILFDCSI-AENIAYGDNSR   GDKGTQLSGGQKQRIAIARALVRQPHILLLDEATSALDTESEK
mmdr2 (N)  PSRANIKILKGLNLKVKSGQTVALVGNSGCGKSTTVQLLQR   IGVVSQEPVLSFTTI-AENIRYGRGNV   GDRGAQLSGGQKQRIAIARALVRNPKILLLDEATSALDTESEA
mmdr2 (C)  PTRANVPVLQGLSLEVKKGQTLALVGSSGCGKSTVVQLLER   LGIVSQEPILFDCSI-AENIAYGDNSR   GDKGTQLSGGQKQRIAIARALIRQPRVLLLDEATSALDTESEK
pfmdr (N)  DTRKDVEIYKDLSFTLLKEGKTYAFVGESGCGKSTILKLIE   IGVVSQDPLLFSNSI-KNNIKYSLYSL   GSNASKLSGGQKQRISIARAIMRNPKILILDEATSSLDNKSEY
pfmdr (C)  ISRPNVPIYKNLSFTCDSKKTTAIVGETGSGKSTFMNLLLR   FSIVSQEPMLFNMSI-YENIKFGREDA   PYGKS-LSGGQKQRIAIARALLREPKILLLDEATSSLDSNSEK
STE6  (N)  PSRPSEAVLKNVSLNFSAGQFTFIVGKSGSGKSTLSNLLLR   ITVVEQRCTLFNDTL-RKNILLGSTDS   GTGGVTLSGGQQQRVAIARAFIRDTPILFLDEAVSALDIVHRN
STE6  (C)  PSAPTAFVYKNMNFDMFCGQTLGIIGESGTGKSTLVLLLTK   ISVVEQKPLLFNGTI-RDNLTYGLQDE   RIDTTLLSGGQAQRLCIARALLRKSKILILDECTSALDSVSSS
hlyB       YKPDSPVILDNINISIKQGEVIGIVGRSGSGKSTLIKLIQR   VGVVLQDNVLLNRSI-IDNISLAPGMS   GEQGAGLSGGQRQRIAIARALVNNPKILIFDEATSALDYASEH
White      IPAPRKHLLKNVCGVAYPGELLAVMGSSGAGKTTLLNALAF   RCAYVQQDDLFIGLIAREHLIFQAMVR   PGRVKGLSGGERKRLAFASEALTDPPLLICDEPTSGLDSFTAH
MbpX       KSLGNLKILDRVSLYVPKFSLIALLGPSGSGKSSLLRILAG   MSFVFQHYALFKHMTVYENISFGLRLR   FEYPAQLSGGQKQRVALARSLAIQPDLLL-DEPFGALDGELRR
BtuD       QDVAESTRLGPLSGEVRAGRILHLVGPNGAGKSTLLARIAG   YLSQQQTPPFATPVWHYLTLHQHDKTR   GRSTNQLSGGEWQRVRLAAVVLQITLLLLDEPMNSLDVAQQSA
PstB       FYYGKFHALKNINLDTAKNQVTAFIGPSGCGKSTLLRTFNK   VGMVFQKPTPFPMSI-YDNIAFGVRLF   HQSGYSLSGGQQQRLCIARGIAIRPEVLLLDEPCSALDPISTG
hisP       RRYGGHEVLKGVSLQARAGDVISIIGSSGSGKSTFLRCINF   GIMVFQHFNLWSHMTVLENVMEAPIQV   GKYPVHLSGGQQQRVSIARALAMEPDVLLFDEPTSALDPELVG
malK       KAWGEVVVSKDINIDIHEGEFVVFVGPSGCGKSTLLRMIAG   VGMVFQSYALYPHLSVAENMSFGLKPA   DRKPKALSGGQRQRVAIGRTLVAEPSVFLLDEPLSNLDAALRV
oppD       TPDGDVTAVNDLNFTLRAGETLGIVGESGSGKSQTAFALMG   ISMIFQDPMTSLNPYMRVGEQLMEVLM   KMYPHEFSGGMRQRVMIAMALLCRPKLLIADEPTTALDVTVQA
oppF       QPPKTLKAVDGVTLRLYEGETLGVVGESGCGKSTFARAIIG   IQMIFQDPLASLNPRMTIGEIIAEPLR   NRYPHEFSGGQCQRIGIARALILEPKLIICDDAVSALDVSIQA
RbsA  (N)  KAVPGVKALSGAALNVYPGRVMALVGENGAGKSTMMKVLTG   AGIIHQELNLIPQLTIAENIFLGREFV   DKLVGDLSIGDQQMVEIAKVLSFESKVIIMDEPTCALIDTETE
RbsA  (C)  VDNLCGPGVNDVSFTLRKGEILGVSGLMGAGRTELMKVLYG   ISEDRKRDGLVLGMSVKENMSLTALRY   EQAIGLLSGGNQQKVAIARGLMTRPKVLILDEPTPGVDVGAKK
UvrA       LTGARGNNLKDVTLTLPVGLFTCITGVSGSGKSTLINDTLF   TYTGVFTPVRELFAGVPESRARGYTPG   GQSATTLSGGEAQRVKLARELSKRGLYILDEPTTGLHFADIQQ
NodI       KSYGGKIVVNDLSFTIAAGECFGLLGPNGAGKSTIIRMILG   IGIVSQEDNLDLEFTVRENLLVYGRYF   NTRVADLSGGMKRRLTLAGALINDPQLLILDEPTTGLDPHARH
FtsE       AYLGGRQALQGVTFHMQPGEMAFLTGHSGAGKSTLLKLICG   IGMIFQDHHLLMDRTVYDNVAIPLIIA   KNFPIQLSGGEQQRVGIARAVVNKPAVLLADEPTGNLDDALSE
```

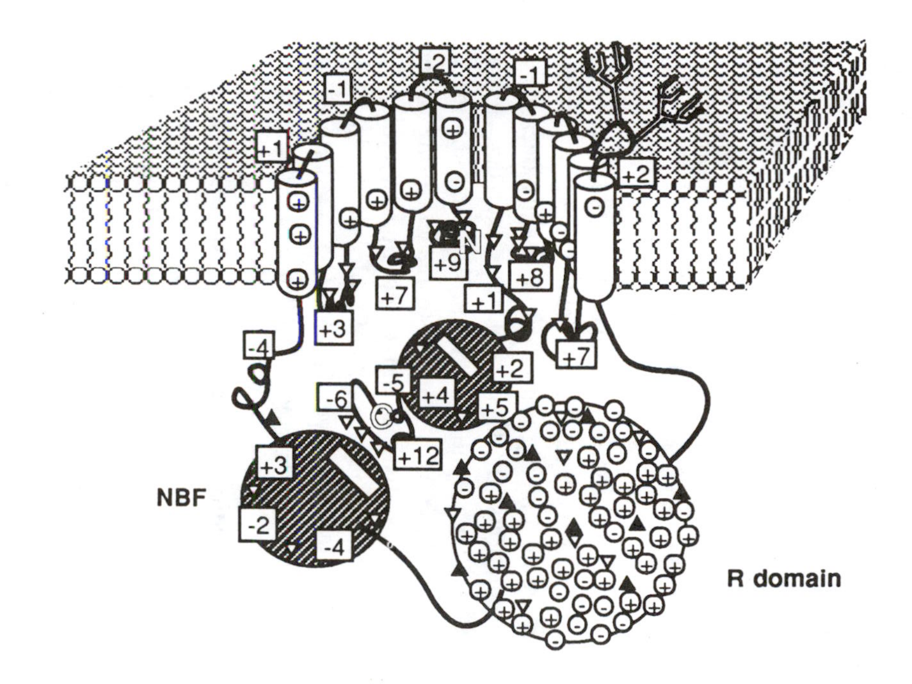

FIGURE 1.12 (A) The protein sequence of the cystic fibrosis gene showed striking similarities to a variety of proteins known to transport molecules across cell membranes. (B) Based on these similarities, it was possible to construct a basic molecular model of the architecture of the CF protein. Reprinted, by permission, from Riordan et al. (1989). Copyright © 1989 by the American Association for the Advancement of Science.

for mathematics, statistics, and computer science in modern molecular biology.

COMING ATTRACTIONS

The chapters of this book describe important applications of mathematical, statistical, and computational methods to molecular biology. These methods are developing rapidly, and, mainly because of this situation, the presentations in this book are intended to be introductory sketches rather than scholarly reviews. Without claiming to be a complete survey, this book should convey to readers some of the exciting uses of mathematics, statistics, and computing in molecular biology. Other introductions to various aspects of molecular biology can be found in Watson et al. (1994), Streyer (1988), U.S. Department of Energy (1992), Watson et al. (1987), Lewin (1990), and Alberts et al. (1989).

Chapter 2 ("Mapping Heredity") describes how statistical models can be used to map the approximate location of genes on chromosomes. Gene mapping was mentioned above for the case of the cystic fibrosis gene. The problem becomes especially challenging—and mathematics plays a bigger role—when the disease does not follow simple Mendelian inheritance patterns—for example, when it is caused by multiple genes or when the trait is quantitative rather than qualitative in nature. This is an important subject for the Human Genome Project and its applications in modern medical genetics.

The next three chapters focus on the analysis of DNA and protein sequences. As new genes are sequenced, they are routinely compared with public databases to look for similarities that might indicate common evolutionary origin, structure, or function. As databases expand at ever-increasing rates, the computational efficiency of such comparisons is crucial. Chapter 3 ("Seeing Conserved Signals") describes combinatorial algorithms for this problem. Because coincidences abound in such comparisons, careful statistical analysis is needed. Chapter 4 ("Hearing Distant Echoes") discusses the application of extremal statistics to sequence similarity. For closely related sequences, sequence comparison also sheds light on the process of evolution. Chapter 5 ("Calibrating the Clock") discusses the applications of stochastic processes to such evolutionary analysis. The discovery and reading of genetic sequences have breathed new life into the study of the stochastic processes of evolution. The chapter focuses on one of the most exciting new tools, the use of the coalescent to estimate times to the most recent common ancestor.

Geometric methods applied to DNA structure and function are the focus of the next three chapters. Watson and Crick's famous DNA double helix can be thought of as local geometrical structure. There is also much interesting geometry in the more global structure of DNA molecules. Chapter 6 ("Winding the Double Helix") uses methods from geometry to describe the coiling and packing of chromosomes. The chapter describes the supercoiling of the double helix, in terms of key geometric quantities—link, twist, and writhe—that are related by a fundamental theorem. Chapter 7 ("Unwinding the Double Helix") employs differential mechanics to study how stresses on a DNA molecule cause it to unwind in certain areas, thereby allowing access by key enzymes needed for gene expression. Chapter 8 ("Lifting the Curtain") uses topology to infer the mechanism of enzymes that recombine DNA strands, providing a glimpse of details that cannot be seen via experiment.

Finally, Chapter 9 ("Folding the Sheets") discusses one of the hardest open questions in computational biology: the protein-folding problem, which concerns predicting the three-dimensional structure of a protein on the basis of the sequence of its amino acids. Probably no simple solution will ever be given for this central problem, but many useful and interesting approximate approaches have been developed. The concluding chapter surveys various computational approaches for structure prediction.

Together, these chapters provide glimpses of the roles of mathematics, statistics, and computing in some of the most exciting and dynamic areas of molecular biology. If this book tempts some mathematicians, statisticians, and computational scientists to learn more about and to contribute to molecular biology, it will have accomplished one of its goals. Its two other goals are to encourage molecular biologists to be more cognizant of the importance of the mathematical and computational sciences in molecular biology and to encourage scientifically literate people to be aware of the increasing impact of both molecular biology and mathematical and computational sciences on their lives. If this book makes progress toward these three goals, it shall have been well worth the effort.

REFERENCES

Alberts, B., D. Bray, J. Lewis, M. Raff, K. Roberts, and J.D. Watson, 1989, *Molecular Biology of the Cell*, 2nd ed., New York: Garland.

Avery, O.T., C.M. McLeod, and M. McCarty, 1944, "Studies on the chemical nature of the substance inducing transformation of pneumococcal types," *J. Exp. Med.* **79**, 137-158.

Lewin, B., 1990, *Genes IV,* Oxford: Oxford University Press.

National Research Council, 1988, *Mapping and Sequencing the Human Genome,* Washington, D.C.: National Academy Press.

National Research Council, 1992, *DNA Technology in Forensic Science*, Washington, D.C.: National Academy Press.

Richardson, J. S., and D. C. Richardson, 1989, "Principles and patterns of protein conformation," pp. 1-98 in *Prediction of Protein Structure and the Principles of Protein Conformation,* Gerald D. Fasman (ed.), New York: Plenum Publishing Corporation.

Riordan, J.R., J.M. Rommens, B. Kreme, N. Alon, R. Rozmahel, Z. Grzelczak, J. Zielenski, S. Lok, N. Plavsic, J-L. Chou, M.L. Drumm, M.C. Innuzzi, F.S. Collins, and L-C. Tsui, 1989, "Identification of the cystic fibrosis gene: cloning and characterization of complementary DNA," *Science* **245** (September 8), 1066-1073.

Sali, A., E. Shakhnovich, and M. Karplus, 1994. "How does a protein fold?" *Nature* **369** (19 May), 248-251.

Streyer, Lubert, 1988, *Biochemistry,* San Francisco, Calif.: W.H. Freeman.

U.S. Department of Energy, Human Genome Program, 1992, *Primer on Molecular Genetics,* Office of Energy Research, Office of Health and Environmental Research, Washington, D.C.: U.S. Government Printing Office.

Watson, J.D., N. Hopkins, J. Roberts, J.A. Steitz, and A. Weiner, 1987, *Molecular Biology of the Gene*, Menlo Park, Calif.: Benjamin-Cummings.

Watson, J.D., J. Tooze, and D.T. Kurtz, 1994. *Recombinant DNA: A Short Course,* 2nd ed., New York: W.H. Freeman and Co.

Chapter 2
Mapping Heredity: Using Probabilistic Models and Algorithms to Map Genes and Genomes

Eric S. Lander
Whitehead Institute for Biomedical Research
and Massachusetts Institute of Technology

For scientists hunting for the genetic basis of inherited diseases, the human genome is a vast place to search. Genetic diseases can involve such subtle alterations as a one-letter misspelling in 3 billion letters of genetic information. To make the task feasible, geneticists narrow down genes in a hierarchical fashion by using various types of maps. Two of the most important maps—genetic maps and physical maps—depend intimately on mathematical and statistical analysis. This chapter describes how the search for disease genes touches on such diverse topics as the extreme behavior of Gaussian diffusion processes and the use of combinatorial algorithms for characterizing graphs.

The human genome is a vast jungle in which to hunt for genes causing inherited diseases. Even a one-letter error in the 3×10^9 base pairs of deoxyribonucleic acid (DNA) inherited from either parent may be sufficient to cause a disease. Thus, to detect inherited diseases, one must be able to detect mistakes present at just over 1 part in 10^{10}. The task is sometimes likened to finding a needle in a haystack, but this analogy actually understates the problem: the typical 2-gram needle in a 6,000-kilogram haystack represents a 1,000-fold larger target. In certain respects, the gene hunter's task is harder still, because it may be difficult to recognize the target even if one stumbles upon it.

The human genome is divided into 23 chromosome pairs, consisting of 1 pair of sex chromosomes (XX or XY) and 22 pairs of autosomes. The number of genes in the 3×10^9 nucleotides of the human DNA sequence is uncertain, although a reasonable guess is 50,000 to 100,000, based on the estimate that a typical gene is about 30,000 nucleotides long. This estimate is only rough, because genes can vary from 200 base pairs to 2×10^6 base pairs in length, and because it is hard to draw a truly random sample.

Although molecular biologists refer to the human genome as if it were well defined in mathematicians' terms, it is recognized that, except for identical twins, no two humans have identical DNA sequences. Two genomes chosen from the human population are about 99.9 percent identical, affirming our common heritage as a species. But the 0.1 percent variation translates into some 3 million sequence differences, pointing to each individual's uniqueness. Common sites of sequence variations are called **DNA polymorphisms**. Most polymorphisms are thought to be functionally unimportant variations—arising by mutation, having no deleterious consequences, and increasing (and decreasing) in frequency by stochastic drift. The presence of considerable DNA polymorphism in the population has sobering consequences for disease hunting. Even if it were straightforward to determine the entire DNA sequence of individuals (in fact, determining a single human sequence is the focus of the entire Human Genome Project), one could not find the gene for cystic fibrosis (CF) simply by comparing the sequences of a CF patient and an unaffected person: there would be too many polymorphisms.

How then does a geneticist find the genes responsible for cystic fibrosis, diabetes, or heart disease? The answer is to proceed hierarchically. The first step is to use a technique called genetic mapping to narrow down the location of the gene to about 1/1,000 of the human genome. The second step is to use a technique called physical mapping to clone the DNA from this region and to use molecular biological tools to identify all the genes. The third step is to identify candidate genes (based on the pattern of gene expression in different tissues and at different times) and look for functional sequence differences in the DNA (for example, mutations that introduce stop codons or that change crucial amino acids in a protein sequence) of affected patients. This chapter focuses on genetic mapping and physical mapping, because it turns out that each intimately involves mathematical analysis.

GENETIC MAPPING

The Concept of Genetic Maps

Genetic mapping is based on the perhaps counterintuitive notion that it is possible to find *where* a gene is without knowing *what* it is. Specifically, it is possible to identify the location of an unknown disease-causing gene by correlating the inheritance pattern of the disease in families with the inheritance pattern of known genetic markers. To understand the foundation of genetic mapping, it is useful to return to the work of Gregor Mendel.

Based on his experiments with peas (see Chapter 1), Mendel concluded that individuals possess two copies, called alleles, of each gene. Mendel's Laws of Inheritance are as follows:

- *First Law*. For any gene, each parent transmits one allele chosen at random to its offspring.
- *Second Law*. For any two genes, the alleles transmitted by a parent are independent (that is, there is no correlation in the alleles transmitted).

Although Mendel's First Law has held up well over the past 130 years, the Second Law turned out to be false in general. Two genes on different chromosomes show no correlation in their inheritance pattern, but genes on the same chromosome typically show correlation.

Consider the backcross in Figure 2.1, showing the inheritance of two genes A and B on the same chromosome. The F_1 individual carries one chromosome with alleles a_1 and b_1 at the two genes and another chromosome with alleles a_2 and b_2. Often, one or the other chromosome is transmitted completely intact to the offspring. If this always happened, the inheritance pattern at the two genes would be completely dependent: a_1 would always be co-inherited with b_1. But the situation is more interesting. Crossing over can occur at random points along the chromosomes, involving an even swap of DNA material. If a crossover occurs between genes A and B, it results in recombination between the genes, producing a chromosome carrying a new combination of alleles: a_1b_2 or a_2b_1. In fact, multiple crossovers can occur along a chromosome;

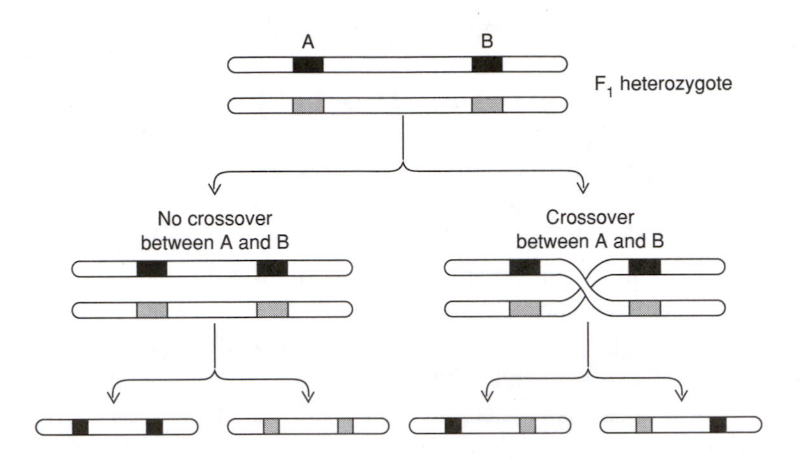

FIGURE 2.1 Schematic drawing of genetic recombination in an F_1 heterozygote with distinct alleles at two loci (marked as A and B) on a chromosome. When no recombination occurs between A and B in meiosis, chromosomes carrying the original pair of alleles result. When recombination occurs, the resulting chromosomes carry a new combination of alleles.

recombination between two loci will result whenever an odd number of crossovers occur.

Genetic mapping is based on the recognition that the recombination frequency θ between two genes (or loci) provides a measure of the distance between them. If two genes are close together, θ will be small. If they lie farther apart, θ will be larger. If the recombination frequency is significantly less than 0.50, the genes are said to be linked. The first genetic linkage map (Figure 2.2), showing the location of six genes in the fruit fly *Drosophila melanogaster*, was constructed in 1911 by Alfred Sturtevant, when he was still a sophomore at Columbia University en route to a career as a great geneticist. Genetic maps were a triumph of abstract mathematical reasoning: Sturtevant was able to chart the location of mutations affecting fly development—even though he understood neither the biochemical basis of the defects nor even that genes were made of DNA!

The genetic distance $d_{\mathbf{A,B}}$ between two genes A and B (measured in units called morgans, after the fly geneticist Thomas Hunt Morgan) is defined as the expected number of crossovers between the genes. If one assumes that crossovers are distributed independently with respect to one

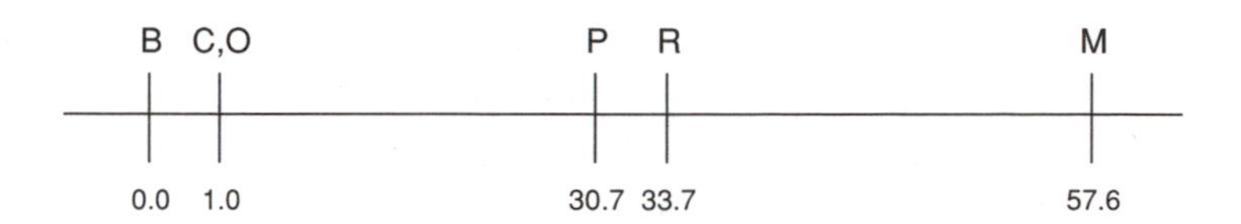

FIGURE 2.2 The first genetic map, showing six loci on the *Drosophila* X chromosome, was constructed by Alfred Sturtevant in 1911.

another, genetic distance can easily be converted into recombination frequency. (This assumption of independence is not quite right but is adequate for many purposes.) For the number of crossovers between genes A and B will then be Poisson distributed with mean $d = d_{\mathbf{A,B}}$, and so the probability of an odd number of crossovers can be shown (by summing alternate terms of the Poisson distribution) to be

$$\theta = \left(1 - e^{-2d}\right)/2.$$

This equation, relating recombination frequency to genetic distance, is known as Haldane's mapping function. For small distances, the formula reduces to $\theta \approx d$, which reflects the fact that the possibility of more than one crossover can be neglected. For large distances *d*, the recombination frequency θ approaches 0.50—that is, independent assortment. Mammalian chromosomes are typically 1 to 3 morgans in length. Geneticists typically report distances between genes in centimorgans.

Incidentally, genetic distance between two genes is not necessarily proportional to the physical distance measured in nucleotides. Since crossing over is a biological process carried out by enzymes acting on the chromosome, the distribution of crossovers need not be (and typically is not) uniform with respect to the DNA sequence. Accordingly, molecular geneticists work with two different kinds of maps: genetic maps based on crossover frequency and physical maps based on nucleotide distances. There can be considerable inhomogeneity between the maps, although human geneticists often employ the rough rule of thumb of 1 centimorgan ≈ 1 megabase. (The relationship between genetic and physical length varies among organisms: 1 centimorgan is about 2 megabases in the mouse, about 200 kilobases in the nematode worm *Caenorhabditis elegans*, and about 3 kilobases in yeast.)

Genetic mapping is an essential first step in characterizing a new mutation causing an interesting phenotype (that is, trait). Consider first the situation of (1) a laboratory organism in which experimental matings can be set up at will and (2) traits that are monogenic and fully penetrant (that is, the phenotype is completely determined by the genotype at a single gene). For example, a *Drosophila* geneticist might find a dominantly acting mutation at a locus A causing flies to have an extra set of wings (in fact, such mutations exist). He would set up crosses with strains carrying different genetic markers (that is, variants in other genes of known location) in order to find the regions showing correlated inheritance. Figure 2.3A shows a backcross of this type. The gene A is clearly not linked to locus B but is tightly linked to locus C. The proportion of recombinant chromosomes provides a straightforward statistical estimator of the recombination frequency. In this case, the recombination frequency between A and C is about 20/200 = 10 percent. The gene A can be positioned more precisely by using the three-point cross shown in Figure 2.3B, in which two nearby genetic markers are segregating. Here, it is clear that A maps about midway between genes C and D (see figure caption).

For experimental organisms and simple traits, genetic mapping provides a straightforward way to locate the trait-causing gene to a small interval—typically about 1 centimorgan or less. In essence, one need only "count recombinants." Because the analysis is so easy, *Drosophila* geneticists rarely need to appeal to statistical or mathematical concepts. For geneticists studying human families or complex traits, however, the situation is quite different.

Challenges of Genetic Mapping: Human Families and Complex Traits

Medical geneticists studying diseases face two major problems: (1) for human diseases, one cannot arrange matings at will but rather must retrospectively interpret existing families; and (2) for both human diseases and animal models of these diseases, the trait may not be simply related to the genotype at a single gene. Owing to these complications, genetic mapping of disease genes often requires sophisticated mathematical analysis.

The first problem is the inability to arrange matings. To offset this limitation, human geneticists need to have a huge collection of frequent, naturally occurring genetic markers so that the inheritance pattern of each chromosomal region can be followed just as if one had deliberately set up a cross incorporating specific genetic markers. Throughout most of the century, only a small number of such naturally occurring genetic markers were known (an example is the ABO blood types), and thus human genetic mapping remained a dormant field. In 1980, David Botstein set off a revolution by recognizing that naturally occurring DNA polymorphisms in the human population filled the need (Botstein et al., 1980). By 1994, over 4,000 DNA polymorphisms had been identified and mapped relative to one another.

Even with a dense genetic map of DNA polymorphisms, human genetic mapping confronts several special problems of incomplete information:

- For individuals homozygous (a_1 / a_1) at a gene, one cannot distinguish at this location between the two homologous chromosomes (that is, the maternally and paternally inherited copies of the chromosome).
- For individuals heterozygous (a_1 / a_2) at a gene, one cannot tell which allele is on the paternal chromosome and which is on the maternal chromosome unless one can study the individual's parents.
- Information for deceased individuals (or for those who choose not to participate in a genetic study) is completely missing from the pedigree. Because of these uncertainties, one often cannot simply count recombinants to estimate recombination frequencies.

Another problem is that many traits and diseases do not follow simple Mendelian rules of inheritance. This problem has several aspects:

- *Incomplete penetrance.* For some "disease genes," the probability that an individual inheriting the disease gene will have the disease phenotype may be less than 1. This probability is called the penetrance of the disease genotype. Penetrance may depend

A

$$\frac{b}{+}\ \frac{c\ a}{+\ +} \times \frac{b}{b}\ \frac{c\ a}{c\ a}$$

$$\downarrow$$

Offspring Genotypes	Number
(1) $\frac{b}{b}\ \frac{c\ a}{c\ a}$	45
(2) $\frac{+}{b}\ \frac{+\ +}{c\ a}$	45
(3) $\frac{+}{b}\ \frac{c\ a}{c\ a}$	45
(4) $\frac{b}{b}\ \frac{+\ +}{c\ a}$	45
(5) $\frac{b}{b}\ \frac{c\ +}{c\ a}$	5
(6) $\frac{+}{b}\ \frac{c\ +}{c\ a}$	5
(7) $\frac{b}{b}\ \frac{+\ a}{c\ a}$	5
(8) $\frac{+}{b}\ \frac{+\ a}{c\ a}$	5

No linkage between B and A:

$$\frac{\text{Recombinant offspring ((3) + (4) + (5) + (8))}}{\text{Total offspring}} = \frac{100}{200} = 50\%$$

Linkage between C and A:

$$\frac{\text{Recombinant offspring ((5) + (6) + (7) + (8))}}{\text{Total offspring}} = \frac{20}{200} = 10\%$$

FIGURE 2.3 Examples of three-point crosses. (A) Locus A is unlinked to locus B but is linked to locus C at a recombination fraction of 10 percent.

B

$$\frac{c\ a\ d}{+\ +\ +} \times \frac{c\ a\ d}{c\ a\ d}$$

↓

Offspring Genotypes	Number
$\frac{c\ a\ d}{c\ a\ d}$	81
$\frac{+\ +\ +}{c\ a\ d}$	81
$\frac{c\ a\ +}{c\ a\ d}$	9
$\frac{+\ +\ d}{c\ a\ d}$	9
$\frac{c\ +\ +}{c\ a\ d}$	9
$\frac{+\ a\ d}{c\ a\ d}$	9
$\frac{c\ +\ d}{c\ a\ d}$	1
$\frac{+\ a\ +}{c\ a\ d}$	1

FIGURE 2.3 (B) Locus A is located between loci C and D, at about 10 percent recombination fraction from each. The first two types of progeny involve chromosomes with no recombination; the next four involve a single recombination, and the last two involve double recombination (between C-A and A-D). The double recombinant class is always least frequent, a property that allows one to determine the order of three linked loci from a cross in which they are all segregating.

on other unknown genes, age, environmental exposure, or random chance. For example, a gene called BRCA1 on chromosome 17 predisposes to early onset of breast cancer in some women, but the penetrance is estimated to be about 60 percent by age 50 and 85 percent by age 80. As a result, one cannot conclude that an unaffected person has inherited a normal copy of the gene.

- *Phenocopy.* Some diseases can be due to nongenetic causes. For example, colon cancer can be caused by mutations in the APC gene on human chromosome 5, but most cases of colon cancer are thought to be nongenetic in origin (and are often attributed to diet). As a result, one cannot conclude that an affected person has necessarily inherited the disease genotype.
- *Genetic heterogeneity.* Some diseases may be caused by mutations in any one of several different genes. Thus, a disease may show linkage to a genetic marker in some families but not in others.
- *Polygenic inheritance.* Some diseases may involve the interaction of mutations at several different genes simultaneously.

Due to the incomplete information on natural families and the uncertainties of complex genetic traits, a human geneticist often cannot reliably infer an individual's genotype based on his or her phenotype; inferences are probabilistic at best. As a result, genetic mapping requires more sophisticated analytical methods than simply counting recombinants between a disease gene and nearby markers.

Animal models of human diseases are slightly simpler, inasmuch as experimental crosses can be arranged. Still, interesting diseases typically show complex inheritance even in inbred animal strains. For example, mouse and rat models of diabetes involve incomplete penetrance, phenocopies, and polygenic inheritance. Sophisticated analytical tools are thus needed for such genetic mapping as well.

MAXIMUM LIKELIHOOD ESTIMATION

To handle the problem of incomplete information, geneticists have adopted the statistical approach of maximum likelihood estimation. Briefly sketched below is the basic formulation (see, e.g., Ott, 1991).

In most cases, a geneticist needs to estimate a parameter θ—for example, the recombination frequency between a disease gene and a genetic marker or the mean increase in blood pressure attributable to a putative gene at a specific location along the chromosome. The geneticist would ideally like to have complete genotypic data X—for example, the genotype for every family member, including the precise parental chromosome from which each allele was inherited. Given complete information, it is usually easy to estimate the required parameters—for example, the recombination frequency can be estimated by counting recombinant chromosomes, and the penetrance can be estimated by finding the proportion of individuals with a disease-predisposing genotype that manifests the disease. Unfortunately, one typically has only incomplete data Y from which it is difficult to estimate θ directly.

The maximum likelihood estimate $\hat{\theta}$ is the value that makes the observed data Y most likely to have occurred, that is, the value that maximizes the likelihood function, $L(\theta){:}=\mathbf{P}(Y|\theta)$. Using Bayes Theorem, one can calculate $L(\theta)$ by the equation:

$$L(\theta){:}=\mathbf{P}(Y|\theta)=\sum_{X}\mathbf{P}(Y|X)\,\mathbf{P}(X|\theta),$$

where the summation is taken over all possible values for the complete data X. The sum is easy to calculate in theory: it decomposes into various terms (corresponding to each individual and each genetic interval) that are conditionally independent, given complete data X.

One can then calculate $\hat{\theta}$ by numerical maximization methods. To determine whether $\hat{\theta}$ is significantly different from a null value θ_0 (for example, to see whether an estimated recombination frequency is significantly less than 50 percent), one examines the likelihood ratio $Z=L(\hat{\theta})/L(\theta_0)$. If Z exceeds some appropriate threshold T, a statistically significant effect has been found.

In principle, virtually any genetic problem can be treated by this approach. In practice, however, two important issues arise: (1) efficient algorithms and (2) statistical significance.

Efficient Algorithms

The number of terms in the sum $L(\theta)$ scales as roughly $O(e^{cmn})$, where m is the number of people in the family, n is the number of genetic markers studied, and c is a constant. Except in the case of the smallest problems, it is infeasible to enumerate all the terms in the sum. Thus, it is a challenge even to calculate the likelihood $L(\theta)$ at a single point, let alone to find the value $\hat{\theta}$ that maximizes the function. Considerable mathematical attention has been devoted to finding efficient ways to calculate $L(\theta)$.

The first breakthrough was due to Elston and Stewart (1971), who devised an algorithm for computing the sum in a nested fashion from the youngest generations of a family to the oldest. The algorithm used the fact that the youngest generation is dependent only on its parents. The running time scales as $O(me^{c'n})$, being exponential only in the number of genetic markers. This algorithm has been a mainstay of genetic calculations but becomes cumbersome when geneticists wish to employ ever-denser maps involving more genetic markers.

Subsequently, Lander and Green (1987) developed an alternative algorithm based on hidden Markov models for nesting the sum according to the genetic markers used, starting with markers at one end of the map and working across to the other end. These authors also exploited the expectation-maximization algorithm from statistics (a well-known maximum likelihood approach) to aid in the multidimensional search for $\hat{\theta}$. The algorithm scales as $O(ne^{c''m})$, being linear in the number of markers although exponential in the number of family members. This algorithm has proven very useful for constructing dense genetic linkage maps, which involves studying many markers in families of limited size. But it is not practical for studying large disease pedigrees.

Recently, mathematical geneticists have explored ways to approximate $L(\theta)$ by sampling from the sum. Modern techniques such as Gibbs sampling and importance sampling have been introduced in the past few years (Kong, 1991; Kong et al., 1992 a,b; Kong et al., 1993; Thompson and Wijsman, 1990). These methods exploit the fact that each piece of missing data depends only on local information in the pedigree: the probability distribution of the genotype of genetic marker i in individual j depends only on the genotypes of nearby markers in nearby individuals. Finding better ways to compute the likelihood function remains a problem from the standpoint of genetics and an excellent test bed for new statistical estimation techniques.

Statistical Significance

In many genetic situations, one may search for a disease gene by estimating $\hat{\theta}$ at many locations along the genome. When multiple comparisons are done, the threshold for statistical significance must be higher than the threshold for a single comparison. But how high should the threshold be? In principle, looking for the presence of a gene at every position along a continuous line involves infinitely many tests—although nearby points are clearly correlated. Surprisingly, the answer to this threshold question turns out to depend on relatively recent results from the theory of large deviations of diffusion processes. This idea is elaborated on in the next section, based on an example from recent work in our laboratory on susceptibility to colon cancer.

Excursion: Susceptibility to Colon Cancer in Mice and the Large Deviation Theory of Diffusion Processes

Colon cancer is one of the most prevalent malignancies in Western societies, with an estimated 145,000 new cases and 60,000 deaths per year in the United States alone. Although environmental factors such as diet can markedly influence the incidence of the disease, genetic factors are known to play a key role. Some families show striking clusters of colon cancer, with aggregations far beyond what could be explained by chance alone. Among such colon cancer families, there is a distinctive subtype called familial adenomatous polyposis (FAP), which is characterized by the fact that affected individuals develop a large number of intestinal growths called polyps that can become tumors. Genetic mapping studies (Bodmer et al., 1987; Leppert et al., 1987) showed that FAP was genetically linked to a region on the long arm of human chromosome 5; subsequently, physical mapping studies led to the isolation of the responsible gene, named APC (Groden et al., 1991; Kinzler et al., 1991; Nishisho et al., 1991).

One way to study the role of APC in tumorigenesis is to turn to biochemistry, in an effort to understand the cellular components with which the protein product interacts. Another way is to turn back to genetics for further insight. One observation about FAP families is that individuals inheriting precisely the same APC mutation may be affected to different degrees. What is the reason for the variability in the manifestation of the disease? Is it due to environment or to the effects of other genes? If the latter, then finding such modifying genes could shed light on the process by which colon cancer develops.

By the usual scientific serendipity, animal studies turned out to hold an important clue. In 1990, William Dove's laboratory at the University of Wisconsin was performing mutagenesis experiments and identified a mouse that spontaneously developed colon tumors (Moser et al., 1990). The dominantly acting mutation responsible for the trait was named Min (for multiple intestinal neoplasia). After considerable genetic mapping and cloning, Dove and his colleagues showed that Min was in fact a mutation in the mouse version of the APC gene (Su et al., 1992).

The Min mouse thus provided a model of human colon cancer and, in particular, a way to look for other genes that might suppress the development of colon tumors. The Min mutation is usually maintained in a heterozygous state on a mouse strain called B6, and such B6 Min/+ mice typically develop about 30 intestinal tumors and die by 3 to 4 months of age. When Dove and his colleagues crossed this mouse to another mouse strain called AKR, they got a surprising result: the F_1 Min/+ progeny developed many fewer colon tumors. On average, the F_1 mice developed about six tumors and most did not die from them. Somehow, the AKR strain must have contributed alleles at one or more genes that substantially modified the effects of Min. Dove's laboratory and our laboratory decided to collaborate to try to map the modifying genes (Dietrich et al., 1993).

A backcross was arranged in which the F_1 progeny were mated back to the more susceptible B6 strain (Figure 2.4). For any modifier locus, 50 percent of the progeny should inherit one copy of the suppressing allele from the AKR strain (that is, have genotype *AB*) and 50 percent should be homozygous for the nonsuppressing allele from B6 (that is, have genotype *BB*). Each animal inheriting the Min mutation was scored for its phenotype by dissecting the intestine and counting the number of tumors and for its genotype by typing the mice for a dense map of DNA polymorphisms that had been constructed in our laboratory (Dietrich et al., 1992).

The complete data for animal *i* consists of two parts: phenotype ϕ_i and a continuous function $g_i(x)$ indicating the genotype—which is either *AB* or *AB*—at each position along the chromosome (Figure 2.5). Actually,

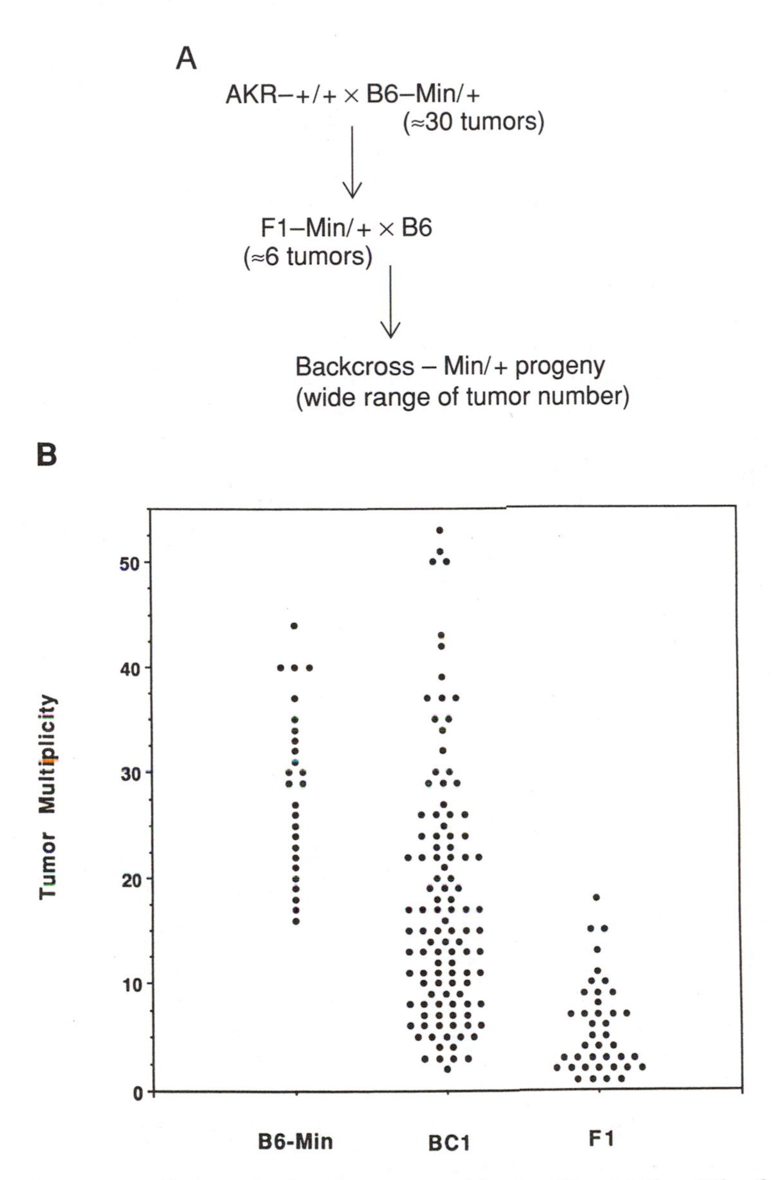

FIGURE 2.4 Distribution of colon tumors caused by the Min mutation. Mice from the B6 strain carrying the genotype Min/+ develop about 30 tumors on average. When these mice are crossed to the AKR strain, the resulting F_1 progeny develop only about 6 tumors. When the F_1 progeny are crossed back to the B6 strain, the resulting backcross progeny show a wide distribution in tumor number. (A) Design of cross. (B) Scatterplot of tumor numbers from different generations in the cross.

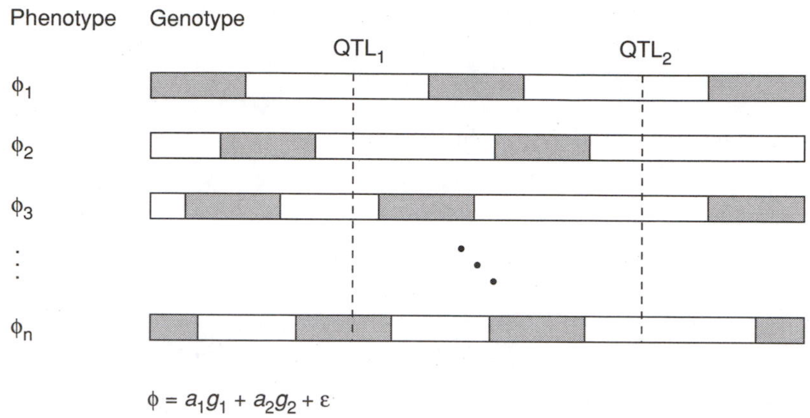

Phenotype Genotype

QTL$_1$ QTL$_2$

ϕ_1

ϕ_2

ϕ_3

$\vdots$

ϕ_n

$\phi = a_1 g_1 + a_2 g_2 + \varepsilon$

FIGURE 2.5 Schematic representation of data for genetic analysis of quantitative traits in a backcross. Every offspring ($i = 1, 2, \ldots, n$) has a phenotype that is a continuous variable ϕ_i and a genotype at every position in the genome. The genotype $g_i(x)$ at position x has two possible states in a backcross (homozygous or heterozygous, encoded as 0 or 1 and represented by black or white in the figure). The figure illustrates the case where the phenotype might depend on two quantitative trait loci (QTL$_1$ and QTL$_2$), according to a linear model $\phi = a_1 g_1 + a_2 g_2 + \varepsilon$, where g_1 is the genotype at QTL$_i$, the a_1 are constants, and ε is a normal random variable.

the problem is slightly more complicated because one can only observe the genotype at the location of the DNA polymorphisms studied. However, for this discussion, the map can be assumed to be so dense that the data are essentially continuous. It can also be assumed that the number n of progeny is very large.

At every position x along the chromosome, the animals can be divided into two sets according to their genotype:

$$AB(x) = \{\text{animal } i \mid g_i(x) = AA\}$$
$$\text{and} \quad BB(x) = \{\text{animal } i \mid g_i(x) = BA\}.$$

If a major modifier gene occurs at location x^*, then the animals in $AB(x^*)$ should have many fewer tumors than the animals in $BB(x^*)$. One could thus perform a t-test (the usual two sample t statistic based on the number of tumors per animal in the two groups) at every position along the chromosome to find a region where the t-statistic Z exceeds some critical threshold T.

How high a threshold is needed to ensure statistical significance, if one scans the entire genome? If for a single chromosome there is no modifying gene along the chromosome, the t-statistic $Z(x)$ at any given point x should be normally distributed with mean 0. It is thus easy to determine the appropriate significance level for the single test at x. But we need to know about the distribution of max $Z(x)$, where the maximum is taken over the entire chromosome.

This question belongs to the field of Gaussian processes. A family of variables $\{Y(x), a \leq x \leq b\}$ is called a Gaussian process if for each $n = 1,2,\ldots$ and each $x_1 < x_2 < \ldots < x_n$, the random variables $Z(x_1), Z(x_2), \ldots, Z(x_n)$ are jointly normally distributed. A Gaussian process is specified by its mean $\mu(t) = \mathbf{E}(Z(t))$ and its covariance $\mathbf{C}(s,t) = \text{cov}(Z(s), Z(t))$. An important example is the "Ornstein-Uhlenbeck process," in which $\mu(t) = 0$ and $\mathbf{C}(s,t) = e^{-\beta|s-t|}$. The Ornstein-Uhlenbeck process arises naturally in physics, because it describes the behavior of a particle undergoing Brownian motion trapped in a potential well. In recent years, Gaussian processes have been a subject of considerable mathematical interest, and the large deviation theory has been worked out for many cases, including the Ornstein-Uhlenbeck process.

Interestingly, it is not hard to show that the statistic $Z(x)$ in our genetic example also follows an Ornstein-Uhlenbeck process with $\beta = 2$. (The mean is 0, and the covariance follows essentially from the Haldane mapping function mentioned above.) Using recent mathematical results (Feingold et al., 1993; Lander and Botstein, 1989), one can thus show that, for large t,

$$\mathbf{P}\{\max_{0 \leq x \leq G} Z(x) \geq t\} \sim (C + 2Gt^2)(1 - \Phi(t)),$$

where Φ (t) is the standard normal cumulative distribution function, C is the number of chromosomes, and G is the length of the genome in morgans. In short, the probability of exceeding threshold t somewhere along a genome of length G is larger by a factor of about $2Gt^2$ than the probability of exceeding it at a single point.

Returning to the problem of colon cancer, we applied this analysis to the mouse genome, which has $C = 20$ chromosomes and genetic length $G \approx 16$. By genetic mapping, we found a striking region on mouse chromosome 4 for which $Z_{max} = 4.3$. The nominal significance level of the statistic is $p = 1.7 \times 10^{-5}$. After correcting for searching over an entire genome (by multiplying by $2G(Z_{max})^2$, the significance level is $p \approx 0.01$. This suggests that there is indeed a modifying gene in this region of chromosome 4.

On the strength of this analysis, several additional crosses were arranged to confirm this result. With more than 300 animals analyzed, the results are now unambiguous: the corrected significance level is now $< 10^{-10}$, and it appears that a single copy of the suppressing form of the gene can decrease tumor number at least twofold. Experiments are now under way to clone the gene, in order to learn its role in reducing colon cancer in genetically predisposed mice. With luck, it may suggest ways to do the same in humans.

PHYSICAL MAPPING

Assembling Physical Maps by "Fingerprinting" Random Clones

Genetic mapping is only the first step toward positional cloning of a gene. Once a gene has been determined to lie between two genetic markers, the geneticist must produce a physical map—consisting of overlapping clones spanning the chromosomal region between the two flanking markers. Traditionally, physical maps have been produced by the process of chromosomal walking: one starts with clone C_1 containing one of the genetic markers, uses C_1 as a probe to find an overlapping clone C_2, uses C_2 as a probe to find C_3, and so on until the region has been spanned (Figure 2.6). Chromosomal walking is an inherently serial procedure, and each step may take several weeks (due to the laboratory procedures involved in making and using a probe).

This tedious process could be eliminated if one simply constructed a complete physical map of overlapping clones spanning the entire genome. The idea is more practical than it may seem at first glance. Whereas chromosomal walking proceeds serially, a physical map of an entire

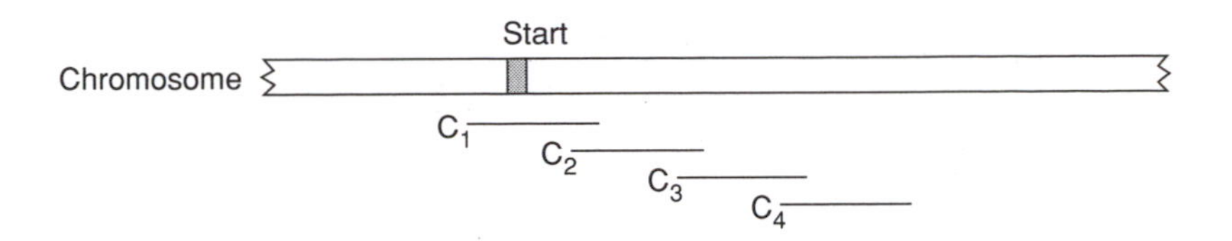

FIGURE 2.6 Schematic diagram illustrating chromosome walking. One starts by isolating a clone C_1 containing the initial starting point. C_1 is then used as a probe to isolate overlapping clones, such as C_2. The process is iterated to obtain successive steps in the walk. Although at each step one isolates clones extending in either direction, only those clones extending the walk to the right are shown in the diagram.

genome can be constructed in parallel. The idea is to describe each clone C by an easily determined fingerprint F(C)—which can be thought of as a set of "attributes" of C. If two clones have substantial overlap, their fingerprints should be similar. Conversely, if two clones have very similar fingerprints, they are likely to overlap. In principle, one should be able to construct a physical map by fingerprinting a large collection of clones and using computer analysis to compare the fingerprints and recognize the overlaps.

The choice of a fingerprinting method depends principally on laboratory considerations; certain types of clones are more amenable to certain types of analysis. Given a large collection of random subclones taken from a genome *G*, possible fingerprints include the following:

- *Complete DNA sequence*. For very small genomes such as those of viruses, it is practical to reassemble the genome from very short subclones of length ~300 to 500 base pairs. For such short subclones, the best fingerprint is the complete DNA sequence of the subclone. It turns out to be relatively easy to sequence such short subclones in one laboratory step, and the resulting sequence provides the most complete possible fingerprint of the clone. Using this information, one can attempt to find the overlaps and piece together the sequence. In fact, this is a widely used technique, referred to as "shotgun" sequencing (Figure 2.7). However, the method is effective only for genomes of length $< 100,000$ base pairs. For larger genomes (such as the genome of even the simplest bacterium), it is difficult to analyze enough subclones to ensure that the entire genome is covered

Complete DNA sequence

```
attgatctcctagtctagttcgatcgggatctcaatcacaccctgcatgttacattgcatacgttagcattacgg
```

Fragments

```
attgatctcctagtctagt
          tagtctagttcgatcgggatctcaatcaca
                    cgatcgggatctcaatcacacc
                             tctcaatcacaccctgcatgtt
                                              atgttacattgcatacgtta
                                                           tacgttagcattacgg
```

FIGURE 2.7 Schematic diagram illustrating "shotgun" DNA sequencing assembly. To obtain the sequence of a larger piece of DNA, one determines the sequence of random subclones and pieces together the complete pieces based on the overlaps. In practice, the subclones are considerably larger than those shown (typically 300 to 500 base pairs) and the overlaps used in assembling the sequence are much larger.

(see the discussion of the coverage problem below). Moreover, the ability to reassemble the sequence is stymied by the frequent occurrence of repeat sequences, which hamper the recognition of overlaps. Nonetheless, shotgun sequencing of small subclones is the method of choice for sequencing moderate-sized DNA fragments.

- *Restriction map.* Larger genomes must be analyzed by studying larger subclones. Such subclones are typically too large to be conveniently sequenced. Instead, restriction maps can provide a useful fingerprint. Restriction maps show the positions of recognition sites at which particular restriction enzymes cut. For example, the restriction enzyme EcoRI cleaves at the sequence `GAATTC`. In effect, a restriction map is an ordered list of the restriction fragments in a clone. To make a restriction map, one can use the method of partial digestion: one radioactively labels one end of a clone, adds a restriction enzyme briefly so that only a random selection of the sites are cut, and measures the lengths of the resulting fragments (Figure 2.8). Restriction maps can be efficiently constructed for clones of moderate size (up to about 50,000 base pairs), although the procedure can be tedious and exacting. If two clones have restriction maps that share several consecutive fragments, it is a good bet that they overlap. With this strategy, Kohara and colleagues (1987) constructed a complete physical map of the bacterium *Escherichia coli* with a genome of 4.6 million base pairs using phage clones containing fragments of about 15,000 base pairs.

Restriction fragment sizes. Rather than constructing an ordered list of the restriction fragments, one can construct an unordered list. This turns out to be technically simpler, because one need not carefully control the rate of cutting as in partial digestion. Clones can instead be digested to completion and the fragment lengths measured. Although the unordered list contains less information, it can still provide an adequate fingerprint. For

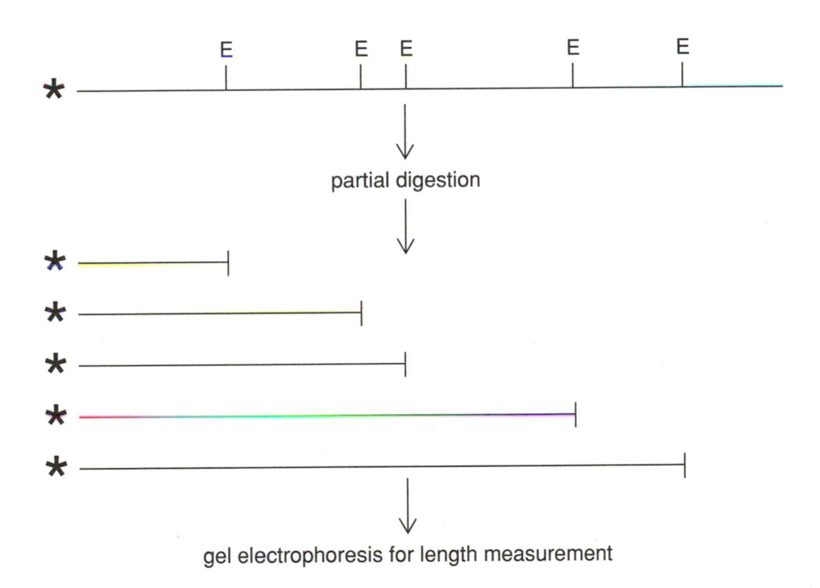

FIGURE 2.8 Schematic diagram illustrating restriction mapping of a DNA fragment by partial digestion. The DNA fragment at the top has several sites (denoted by E) that can be cleaved by the restriction enzyme EcoRI. A large collection of molecules of this DNA fragment is radioactively labeled at one end (denoted by a star) and then exposed briefly to the restriction enzyme. The period of exposure is sufficiently brief that the enzyme can cleave only about one site per molecule, resulting in a collection of radioactively labeled fragments terminating at the various E sites. The length of these fragments (and thus the positions of the E sites) can be determined by gel electrophoresis of the fragments and subsequent exposure of the gel to x-ray film.

example, Olson and colleagues (1986) used this approach to construct a physical map of the yeast *Saccharomyces cerevisiae* with a genome of 13 million base pairs.

- *Content of sequence tagged sites*. For very large genomes such as the human genome with 3×10^9 base pairs, it is necessary to work with large subclones of length $>100,000$ base pairs. For such large subclones, a different fingerprinting strategy has gained favor in recent years. The method is based on sequence tagged sites (STSs), which are very short unique sequences taken from the genome which can be easily assayed by the polymerase chain reaction (PCR). The fingerprint of a clone is the list of STSs contained within it; the data form an incidence matrix of clones by STSs (Figure 2.9). Clones containing even a single unique STS in common should overlap. As an aside, the determination of which clones contain a given STS is typically made using a combinatorial pool scheme that avoids having to test each STS against each clone (Green and Olson, 1990). Using this approach, Foote et al. (1992) and Chumakov et al. (1992) constructed the first complete maps of human chromosomes (Y and 21, respectively).

Regardless of the experimental details of the fingerprinting scheme, there are two key mathematical issues pertinent to the construction of a physical map:

1. *Algorithms for map assembly*. Given the fingerprinting data, what algorithm should be used for constructing a physical map? This question is closely related to graph theory: given information about adjacency among clones inferred from their fingerprints, one must reconstruct the underlying geometry of the physical map.
2. *Statistics of coverage*. How many clones must be studied to yield a map covering virtually the entire genome? This question belongs to probability theory: assuming that subclones are distributed randomly across the genome, one needs to know the distribution of gaps—uncovered regions or undetected overlaps—in the map.

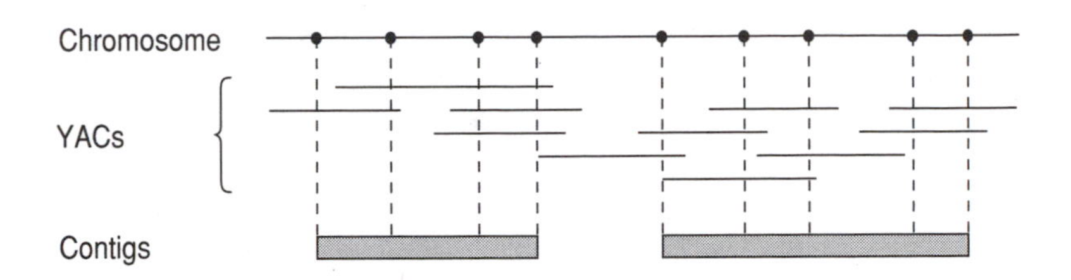

FIGURE 2.9 Schematic diagram illustrating the principle of STS content mapping. Various unique points in the genome, called STSs, are tested against a collection of random large-insert clones, such as YACs, to determine which STSs are contained in which YACs. Based on the resulting adjacency matrix, one attempts to reconstruct the order of the STSs in the genome. "Contigs," consisting of groups of STSs connected by YACs, are assembled based on the adjacency data. In the figure, the STSs can be grouped into two contigs.

Mathematical analysis is thus essential to the design and execution of physical mapping projects (Arratia et al., 1991; Lander and Waterman, 1988). This is illustrated in a discussion below of the considerations involved in making a physical map of the entire human genome.

Excursion: Designing a Strategy to Map the Human Genome

Under the auspices of the Human Genome Project, our laboratory is engaged in constructing complete physical maps of the mouse and human genomes, each about 3×10^9 base pairs in length. The task is daunting, requiring analysis of tens of thousands of clones, each carrying extremely large DNA fragments. Before undertaking such a project, it was crucial to perform careful analysis to identify the best strategy.

Currently, the best clones for making a human physical map are yeast artificial chromosomes (YACs). A good YAC library might contain inserts of about 1 million base pairs in length. Even with such large inserts, it would take 3,000 YACs to cover the human genome if they were laid end-to-end. Of course, clones taken from an actual library will be arrayed randomly, and so considerably more clones are required to ensure coverage.

As noted above, the best fingerprint for studying YACs is STS content mapping. Each STS is screened simultaneously against the entire YAC library to identify the clones that contain it. Because STSs are screened in parallel, it is most efficient to work with a fixed YAC library and to test STSs sequentially.

For mathematical analysis of physical mapping, the YACs and STSs can be abstracted to a set I of intervals (which may vary in size) and a set P of points distributed randomly along a line segment. An interval is said to be anchored if it contains at least one point $p \in P$. Two anchored intervals I_1 and I_2 are said to be connected if there is a point $p \in P$ contained in their intersection. Note that two intervals may overlap but fail to be connected. If we take the transitive closure of the connectivity relation, the resulting equivalence classes of anchored intervals are called anchored "contigs." (For the purpose of the exposition, a definition is used that differs slightly from that in Arratia et al. (1991), in which contigs refer only to equivalence classes containing at least two intervals.)

The key question is: How many intervals and how many points should be analyzed to construct a reasonably complete physical map—that is, one in which the vast majority of the genome is contained in a modest number of large contigs? We define the following notation:

G, the length of the genome in base pairs;
L, the length of a random clone in base pairs, a random variable;
$\mathbf{L}$, the expected length of a random clone, $\mathbf{L} = \mathbf{E}(L)$;
N, the number of clones to be used;
M, the number of STSs to be used;
$a = \mathbf{L}N/G$, the expected number of clones covering a random STS; and
$b = \mathbf{L}M/G$, the expected number of STSs contained in a random clone.

Clone lengths L will be assumed to be independent, identically distributed random variables, with the probability density function of the normalized length $l = L/\mathbf{L}$ denoted by $f(l)$ and the inverse cumulative distribution function (also called the survival function) denoted $F(l) = \mathbf{P}(l/\mathbf{L} > x)$. It is also useful to define the auxiliary function

$$J(x) = \exp\{-a \int_x^\infty F(l)\, dl\},$$

which can be interpreted as the probability that two points separated by distance x are not covered by a common clone.

The problem belongs to the area of coverage problems, which treat processes of covering a space with random sets of a given sort. Often, mathematical authors focus on the goal of attaining *complete* coverage. Such results are not really appropriate from a biological standpoint—because they depend sensitively on the distribution of covering sets being absolutely random, an assumption that is biologically implausible. Instead, it is more sensible to focus on central behavior—that is, the goal of covering *most* of the space.

STS content mapping poses a slightly unusual coverage problem, because the definition of coverage involves joining together random intervals with random points. It is nonetheless possible to analyze many features of the stochastic process in order to derive many prescriptive results. Arratia and colleagues (1991) proved the following result, which describes the basic coverage properties:

Proposition: With the notation as above,

(1) the expected number of anchored contigs is Np_1, where

$$p_1 = \int_x^{\infty} be^{-bu} J(u)F(u)du,$$

(2) the expected length of an anchored contig is $\lambda\,\mathbf{E}(L)$, where

$$\lambda = \{1 + \int_0^{\infty} (b^2u - 2b)e^{-bu} J(u)du\} \,/\, ap_1, \text{ and}$$

(3) the expected proportion r_0 of the genome not covered by anchored contigs is

$$r_0 = \int_0^{\infty}\int_0^{\infty} b^2 e^{-b(u+v)} \frac{J(u)J(v)}{J(u+v)} dudv.$$

Figure 2.10, taken from Arratia et al. (1991), plots these functions for the case of clones of constant size. From these graphs, experimentalists can plan their experimental approach. For our own physical mapping project in the human genome, the typical clone size is about 1×10^6 base pairs. Based on the trade-offs between screening more YACs and using more STSs, we selected $a = 6$ and $b = 3$—corresponding to about 18,000 YACs and about 9,000 STSs. This selection should ensure that about $r_0 \approx 99$ percent of the genome is covered, with about 850 anchored contigs having average length of about 3.5 megabases.

Having explored the question of experimental design, it is worth briefly discussing the issues involved in data analysis. The process of STS content mapping may consume several person-years of laboratory work, but the final result will simply consist of a large (18,000 × 9,000) adjacency matrix $A = (a_{ij})$, with $a_{ij} = 1$ or 0 in position i, j according to whether YAC_i contains STS_j. Based on this information, how do we determine the correct order of the STSs in the genome?

In principle, a proposed order of the STSs is consistent with the observed data if and only if permuting the columns of the adjacency matrix *A* according to this order causes *A* to have the consecutive ones property—that is, in each row, the ones occur in a single consecutive block. This property follows from the fact that each YAC should consist of a single connected interval taken from the genome (see Figure 2.9). The consecutive ones property has been extensively studied in computer science. Booth and Leuker (1975) devised an elegant linear-time algorithm for solving the problem in a very strong sense: Given a (0,1)-matrix *A* with *n* rows and *m* nonzero entries, the algorithm needs a running time of only $O(m+n)$ to determine whether there is *any* column permutation causing the matrix to have the consecutive ones property and, if so, to produce a simple representation of *all* such column permutations.

In practice, there is a serious problem with this approach: it assumes that the data are absolutely error-free. However, laboratory work is never flawless and certainly not when the task involves filling in 162 million entries in an adjacency matrix. If even a few errors are present, the Booth-Leuker algorithm is almost certain to report that there is no consistent order! In fact, there are likely to be many errors, including

- False negatives: one may fail to identify some proportion of the YACs containing an STS;
- False positives: some proportion of the YACs detected as containing an STS may not actually do so; and
- Chimeric YACs: some proportion of the YACs may not represent a single contiguous region, but two unrelated regions that have been joined together in a single clone.

Moreover, the occurrence of false negatives and positives may not be random but systematic (owing to deletions of clones or contamination of samples). In short, algorithms must be robust to errors in the data. Producing such algorithms is an interesting challenge that draws on methods from graph theory, operations research, and statistics. As of this writing, the best approach has not yet been determined.

CONCLUSION

Genetic and physical mapping are key tools for describing the function and structure of chromosomes. Only in the simplest cases is such mapping completely devoid of mathematical issues. In the case of human genetics, mathematics plays a crucial role.

In essence, mapping problems—like many problems in computational biology—involve indirect inference of the structure of a biological entity, such as a chromosome, based on whatever data can be effectively gathered in the laboratory. It is not surprising that mapping problems draw on statistics, probability, and combinatorics. Although the field of mapping dates nearly to the beginning of the 20th century, the area remains rich with new challenges—because new laboratory methods constantly push back the frontiers of the maps and features that can be mapped in DNA.

A

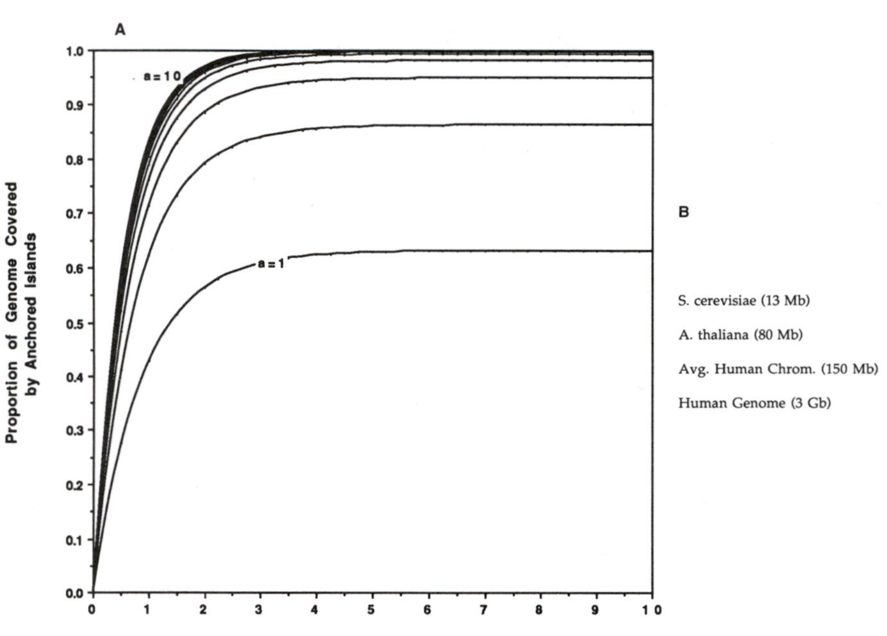

B

Approximate Values of G/L

	Phage (15 kb)	Cosmid (40 kb)	YAC (300 kb)	YAC (1 Mb)
S. cerevisiae (13 Mb)	867	325	43	13
A. thaliana (80 Mb)	5333	2000	267	20
Avg. Human Chrom. (150 Mb)	10,000	3,750	500	150
Human Genome (3 Gb)	200,000	75,000	10,000	3,000

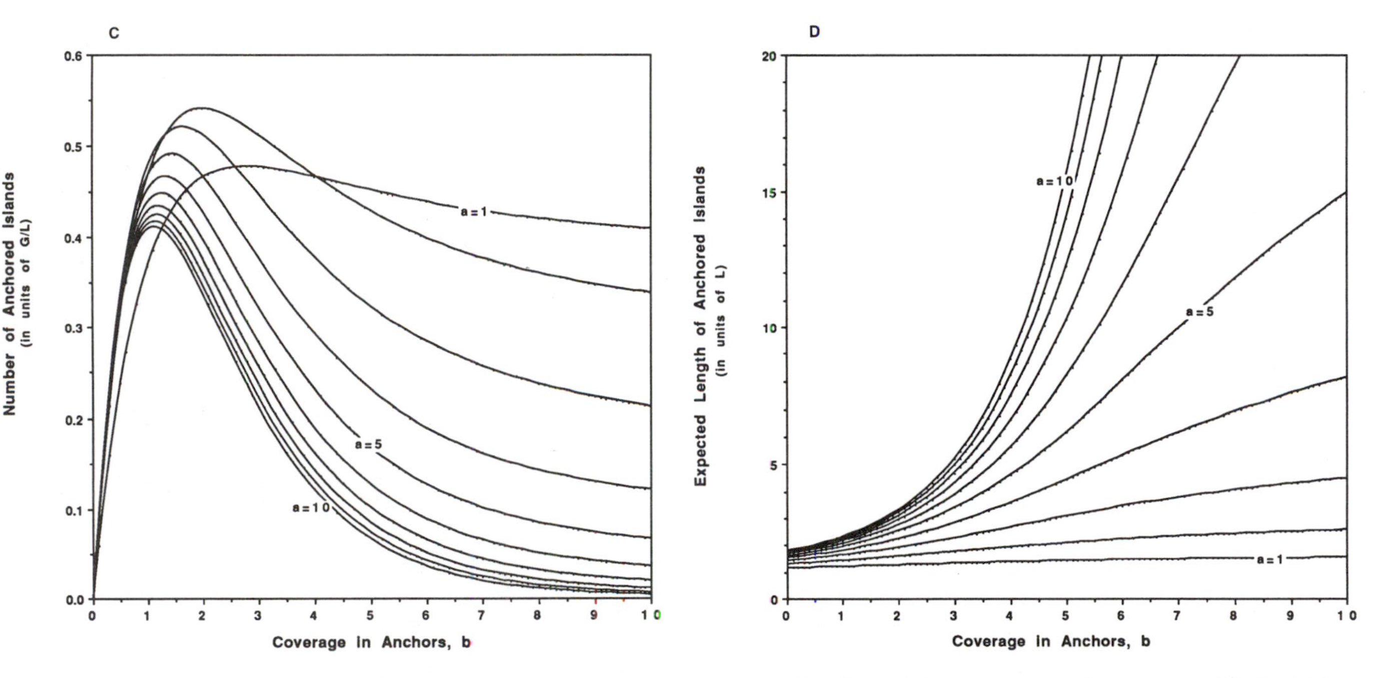

FIGURE 2.10 Expected coverage properties for STS content mapping, as a function of the coverage a in YACs and b in STSs. Calculations assume YACs of constant length L and a genome of length G. The graphs show (A) the expected proportion of the genome covered by anchored "contigs"; (C) the expected number of anchored contigs, and (D) the expected length of an anchored contig. Graphs show the situation for $a = 1,2,\ldots,10$ (only the cases $a = 1,5,10$ are explicitly marked). Results are expressed in units of G/L. Table B lists the value of G/L for certain representative genomes and cloning vectors, including two different sizes of YACs. Reprinted, by permission, from Arratia et al. (1991). Copyright © 1991 by *Genomics*.

REFERENCES

Arratia, R., E.S. Lander, S. Tavaré, and M.S. Waterman, 1991, "Genomic mapping by anchoring random clones: A mathematical analysis," *Genomics* **11**, 806-827.

Bodmer, W.F., C.J. Bailey, J. Bodmer, H.J.R. Bussey, A. Ellis, P. Gorman, F.C. Lucibello, V.A. Murday, S.H. Rider, P. Scambler, D. Sheer, E. Solomon, and N.K. Spurr, 1987, "Localization of the gene for familial adenomatous polyposis on chromosome 5," *Nature* **328**, 614-616.

Booth, K.S., and G.S. Leuker, 1975, "Testing the consecutive ones property, interval graphs, and graph planarity using PQ-tree algorithms," *Journal of Computational Systems Science* **13**, 335-379.

Botstein, D., R.L. White, M. Skolnick, and R.W. Davis, 1980, "Construction of a genetic linkage map in man using restriction fragment length polymorphisms," *American Journal of Human Genetics* **32**, 314-331.

Chumakov, I., P. Rigault, S. Guillou, P. Ougen, A. Billaut, G. Guasconi, P. Gervy, I. Le Gall, P. Soularue, and L. Grinas, 1992, "Continuum of overlapping clones spanning the entire human chromosome 21q," *Nature* **359**, 380-387.

Dietrich, W., H. Katz, S.E. Lincoln, H.-S. Shin, J. Friedman, N.C. Dracopoli, and E.S. Lander, 1992, "A genetic map of the mouse suitable for typing intraspecific crosses," *Genetics* **131**, 423-447.

Dietrich, W.F., E.S. Lander, J.S. Smith, A.R. Moser, K.A. Gould, C. Luongo, N. Borenstein, and W. Dove, 1993, "Genetic identification of Mom-1, a major modifier locus affecting Min-induced intestinal neoplasia in the mouse," *Cell* **75**, 631-639.

Elston, R.C., and J. Stewart, 1971, "A general model for the analysis of pedigree data," *Human Heredity* **21**, 523-542.

Feingold, E., P.O. Brown, and D. Siegmund, 1993, "Gaussian models for genetic linkage analysis using complete high resolution maps of identity by descent," *American Journal of Human Genetics* **53**, 234-251.

Foote, S., D. Vollrath, A. Hilton, and D.C. Page, 1992, "The human Y chromosome: Overlapping DNA clones spanning the euchromatic region," *Science* **258**, 60-66.

Green, E., and M.V. Olson, 1990, "Systematic screening of yeast artificial chromosome libraries using the polymerase chain reaction," *Proceedings of the National Academy of Sciences USA* **87**, 1213-1217.

Groden, J., A. Thliveris, W. Samowitz, M. Carlson, L. Gelbert, H. Albertsen, G. Joslyn, J. Stevens, L. Spirio, M. Robertson, L. Sergeant, K. Krapcho, E. Wolff, R. Burt, J.P. Hughes, J. Warrington, J. McPherson, J. Wasmuth, D. Le Paslier, H. Abderrahim, D. Cohen, M. Leppert, and R. White, 1991, "Identification and characterization of the familial adenomatous polyposis coli gene," *Cell* **66**, 589-600.

Kinzler, K.W., M.C. Nilbert, L.K. Su, B. Vogelstein, T.M. Bryan, D.B. Levy, K.J. Smith, A.C. Preisinger, P. Hedge, D. McKechnie, R. Finniear, A. Markham, J. Groffen, M.S. Boguski, S.F. Altschul, A. Horii, H. Ando, Y. Miyoshi, Y. Miki, I. Nishisho, and Y. Nakamura, 1991, "Identification of FAP locus genes from chromosome 5q21," *Science* **253**, 661-665.

Kohara, Y., A. Akiyama, and K. Isono, 1987, "The physical map of the whole *E. coli* chromosome: Applications of a new strategy for rapid analysis and sorting of a large genomic library," *Cell* **50**, 495-508.

Kong, A., 1991, "Efficient methods for computing linkage of recessive diseases in inbred pedigrees," *Genetics and Epidemiology* **8**, 81-103.

Kong, A., M. Frigge, N. Cox, and W.H. Wong, 1992a, "Linkage analysis with adjustments for covariates: A method combining peeling with Gibbs sampling," *Cytogenetics and Cell Genetics* **59**, 208-210.

Kong, A., M. Frigge, M. Irwin, and N. Cox, 1992b, "Importance sampling. I. Computing multimodel *p* values in linkage analysis," *American Journal of Human Genetics* **51**, 1413-1429.

Kong, A., N. Cox, M. Frigge, and M. Irwin, 1993, "Sequential imputation for multipoint linkage analysis," *Genetics and Epidemiology* **10**, 483-488.

Lander, E.S., and D. Botstein, 1989, "Mapping Mendelian factors underlying quantitative traits using RFLP linkage maps," *Genetics* **121**, 185-199.

Lander, E.S., and P. Green, 1987, "Construction of multilocus genetic linkage maps in humans," *Proceedings of the National Academy of Sciences USA* **84**, 2363-2367.

Lander, E.S., and M.S. Waterman, 1988, "Genomic mapping by fingerprinting random clones: A mathematical analysis," *Genomics* **2**, 231-239.

Leppert, M., M. Dobbs, P. Scambler, P. O'Connell, Y. Nakamura, D. Stauffer, S. Woodward, R. Burt, J. Hughes, E. Gardner, M. Lathrop, J. Wasmuth, J.M. Lalouel, and R. White, 1987, "The gene for familial polyposis coli maps to the long arm of chromosome 5," *Science* **238**, 1411-1413.

Moser, A.R., H.C. Pitot, and W.F. Dove, 1990, "A dominant mutation that predisposes to multiple intestinal neoplasia in the mouse," *Science* **247**, 322-324.

Nishisho, I., Y. Nakamura, Y. Miyoshi, Y. Miki, H. Ando, A. Horii, K. Koyama, J. Utsunomiya, S. Baba, P. Hedge, A. Markham, A.J. Krush, G. Peterson, S.R. Hamilton, M.C. Nilbert, D.B. Levy, T.M. Bryan, A.C. Preisinger, K.J. Smith, L.K. Su, K.W. Kinzler, and B. Vogelstein, 1991, "Mutations of chromosome 5q21 genes in FAP and colorectal cancer patients," *Science* **253**, 665-669.

Olson, M.V., J.E. Dutchik, M.Y. Graham, G.M. Brodeur, C. Helms, M. Frank, M. MacCollin, R. Scheinman, and T. Frank, 1986, "Random-clone strategy for genomic restriction mapping in yeast," *Proceedings of the National Academy of Sciences USA* **83,** 7826-7830.

Ott, J., 1991, *Analysis of Human Genetic Linkage,* Baltimore, Md.: Johns Hopkins University Press.

Su, L.K., K.W. Kinzler, B. Vogelstein, A.C. Preisinger, A.R. Moser, C. Luongo, K.A. Gould, and W.F. Dove, 1992, "A germline mutation of the murine homolog of the APC gene causes multiple intestinal neoplasia," *Science* **256**, 668-670.

Thompson, E., and E. Wijsman, 1990, *The Gibbs Sampler on Extended Pedigrees: Monte Carlo Methods for the Genetic Analysis of Complex Traits*, Technical Report **193,** Department of Statistics, University of Washington, Seattle.

Chapter 3
Seeing Conserved Signals: Using Algorithms to Detect Similarities Between Biosequences

Eugene W. Myers
University of Arizona

The sequence of amino acids in a protein determines its three-dimensional shape, which in turn confers its function. Segments of the protein that are critical to its function resist evolutionary pressures because mutations of such segments are often lethal to the organism. These critical "active sites" tend to be conserved over time and so can be found in many organisms and proteins that have similar function. Analogously, functionally important segments of an organism's DNA tend to be conserved and to recur as common motifs. In this chapter, the author introduces algorithms for comparing DNA and protein sequences to reveal similar regions. Particular attention is given to the problem of searching a large database of catalogued sequences for regions similar to a newly determined sequence of unknown function.

Since the advent of deoxyribonucleic acid (DNA) sequencing technologies in the late 1970s, the amount of data about the protein and DNA sequence of humans and other organisms has been growing at an exponential rate. It is estimated that by the turn of the century there will be terabytes of such biosequence information, including DNA sequences of entire human chromosomes. Databases of these sequences will contain a wealth of information about the nature of life at the molecular level *if* we can decipher their meaning.

Proteins and DNA sequences are polymers consisting of a chain of monomers with a common backbone substructure that links them together. In the case of DNA, there are 4 types of monomers, the nucleotides, each having a different side chain. For proteins, there are 20 types of monomers, the amino acids. With just a few exceptions, the sequence of monomers, that is, the primary structure, of a given protein or DNA strand completely determines the three-dimensional shape of the biopolymer. Because the function of a molecule is determined by the position of its atoms in space, this almost perfect correlation between sequence and structure implies that to know the function of a biopolymer, it in principle suffices to know its primary sequence.

The primary sequence of a DNA segment is denoted by a string consisting of the four letters `A`, `C`, `G`, and `T`. Analogously, the primary sequence of a protein is denoted by a string consisting of 20 letters of the alphabet, one for each type of amino acid. In principle, these strings of symbols encode everything one needs to know about the protein or DNA strand in question. If the primary sequences of two proteins are similar, then it is reasonable to conjecture that they perform the same function. Because DNA's principal role is one of encoding information (including all of an organism's proteins), the similarity of two segments of DNA suggests that they code similar things.

Mutation in a DNA or protein sequence is a natural evolutionary process. Errors in the replication of DNA can cause a change in the nucleotide at a given position. Less often, a nucleotide is deleted or inserted. If the mutation occurs in a region of DNA that codes for protein, these changes cause related changes in the primary sequence and, hence, the shape and activity of the protein. The impact of a particular mutation depends on the degree to which the original and new amino acid sequences differ in their physical and chemical properties. Mutations that result in proteins that are so altered that they function improperly or not at all tend to be lethal to the organism. Nature is biased against mutations in those critical regions central to a protein's function and is more lenient toward changes in other regions.

Similarity of DNA sequences is a clue to common evolutionary origin. If two proteins in two organisms evolved from a common precursor, one will generally find highly similar segments, reflecting strongly conserved critical regions. If the proteins are very recent derivatives, one might expect to see similarity over the entire length of the sequences. While proteins can be similar because of evolution from a common precursor, similarity of protein sequences can also be a clue to common function independent of evolutionary considerations. It appears that nature not only conserves the critical parts of a protein's conformation and function, but also reuses such motifs as modular units in fashioning the spectrum of known proteins. One finds strong similarities between segments of proteins that have similar functions. A strong similarity between the *v-sis* oncogene and a growth-stimulating hormone was the key to discovering that the *v-sis* oncogene causes cancer by deregulating cell growth. In that case, the similarity involved the entirety of the sequence. In other cases, functionally related proteins are similar only in segments corresponding to active sites or other functionally critical stretches.

FINDING GLOBAL SIMILARITIES

To illustrate the underlying techniques of sequence comparison, we begin with a simple, core problem of finding the best alignment between the entirety of two sequences. Such an alignment is called a *global alignment* because it aligns the entire sequences, as opposed to a *local alignment,* which aligns portions of the sequences.

As an example, consider finding the best global alignment of A = `ATTACG` and B = `ATATCG` under the following scoring scheme. A letter aligned with the same letter has a score of 1. A letter aligned with any different letter or a gap has a score of 0. The total score is the sum of the scores for the alignment. A matrix depicting this "unit-cost" scoring scheme is shown in Figure 3.1. Under this unit-cost scheme, the score of an alignment is equal to the number of identical aligned characters. The obvious alignment $\genfrac{}{}{0pt}{}{\texttt{ATTACG}}{\texttt{ATATCG}}$ has a score of 4. However, because gaps are allowed, a higher score can be achieved, namely, 5, which can be shown to be the highest score possible. An optimal alignment, that is, an alignment that achieves this highest score by aligning five symbols, is $\genfrac{}{}{0pt}{}{\texttt{ATTA-CG}}{\texttt{A-TATCG}}$. In some cases, there is only one, unique optimal alignment, but in general there can be many. For example, $\genfrac{}{}{0pt}{}{\texttt{AT-TACG}}{\texttt{ATAT-CG}}$ also has a score of 5.

The unit-cost scoring scheme of Figure 3.1 is not the only possible scheme. Later in this chapter, we will see a much more complex scoring scheme used in the comparison of proteins (20-letter alphabet). In that scheme and other scoring schemes, the scores in the table are real numbers assigned on the basis of various interpretations of empirical evidence. Let us introduce here a formal framework to assist our thinking.

δ	–	A	C	G	T
–	0	0	0	0	0
A	0	1	0	0	0
C	0	0	1	0	0
G	0	0	0	1	0
T	0	0	0	0	1

FIGURE 3.1 Unit-cost scoring scheme.

Consider comparing sequence $\mathbf{A} = a_1 a_2 \cdots a_M$ and sequence $\mathbf{B} = b_1 b_2 \cdots b_N$, whose symbols range over some alphabet Ψ, for example, $\Psi = \{\texttt{A,C,G,T}\}$ for DNA sequences. Let $\delta(a,b)$ be the score for aligning a with b, let $\delta(a,-)$ be the score of leaving symbol a unaligned in sequence **A**, and let $\delta(-,b)$ be the score of leaving b unaligned in **B**. Here a and b range over the symbols in Ψ and the gap symbol "–". The score of an alignment is simply the sum of the scores δ assigns to each pair of aligned symbols, for example, the score of $\begin{matrix}\texttt{ATTA-CG}\\\texttt{A-TATCG}\end{matrix}$ is

$$\delta(\texttt{A},\texttt{A}) + \delta(\texttt{T},-) + \delta(\texttt{T},\texttt{T}) + \delta(\texttt{A},\texttt{A}) + \delta(-,\texttt{T}) + \delta(\texttt{C},\texttt{C}) + \delta(\texttt{G},\texttt{G}),$$

which for the scoring scheme of Figure 3.1 equals 5. An optimal alignment under a given scoring scheme is an alignment that yields the highest sum.

Visualizing Alignments: Edit Graphs

Many investigators have found it illuminating to convert the problem of finding similarities into one of finding certain paths in an *edit graph*.

Proceeding formally, the edit graph $G_{\mathbf{A},\mathbf{B}}$ for comparing sequences **A** and **B** is an edge-labeled directed graph, as illustrated in Figure 3.2 for the example mentioned above. The vertices of the graph are arranged in an $M+1$ by $N+1$ rectangular grid or matrix, so that (i, j) designates the vertex in column *i* and row *j* (where the numbering starts at 0). The following edges, and only these edges, are in $G_{\mathbf{A},\mathbf{B}}$:

1. If $i \in [1, M]$ and $j \in [0, N]$, then there is an **A**-*gap* edge $(i-1, j) \to (i, j)$ labeled $"{a_i \atop -}"$ whose score is $\delta(a_i, -)$.
2. If $i \in [0, M]$ and $j \in [1, N]$, then there is a **B**-*gap* edge $(i, j-1) \to (i, j)$ labeled $"{- \atop b_j}"$ whose score is $\delta(-, b_j)$.
3. If $i \in [1, M]$ and $j \in [1, N]$, then there is an *alignment* edge $(i-1, j-1) \to (i, j)$ labeled $"{a_i \atop b_j}"$ whose score is $\delta(a_i, b_j)$.

The edit graph has the property that paths and alignments between segments of **A** and **B** are in isomorphic correspondence. That is, any path from vertex (g, h) to vertex (i, j) for $g \le i$ and $h \le j$ models an alignment between the substrings $a_{g+1}a_{g+2}\cdots a_i$ and $b_{h+1}b_{h+2}\cdots b_j$, and vice versa. The alignment modeled by a path is the sequence of aligned pairs given by labels on its edges. For example, in Figure 3.2 the two highlighted paths, both from vertex $(0,0)$ to $(6,6)$ correspond to the two optimal global alignments $\begin{smallmatrix}\texttt{ATTA-CG}\\ \texttt{A-TATCG}\end{smallmatrix}$ and $\begin{smallmatrix}\texttt{AT-TACG}\\ \texttt{ATAT-CG}\end{smallmatrix}$.

The Basic Dynamic Programming Algorithm

We now turn to devising an algorithm, or computational procedure, for finding the score of an optimal global alignment between sequences **A** and **B**. We focus on computing just the score for the moment, and return to the goal of delivering an alignment achieving that score at the end of this

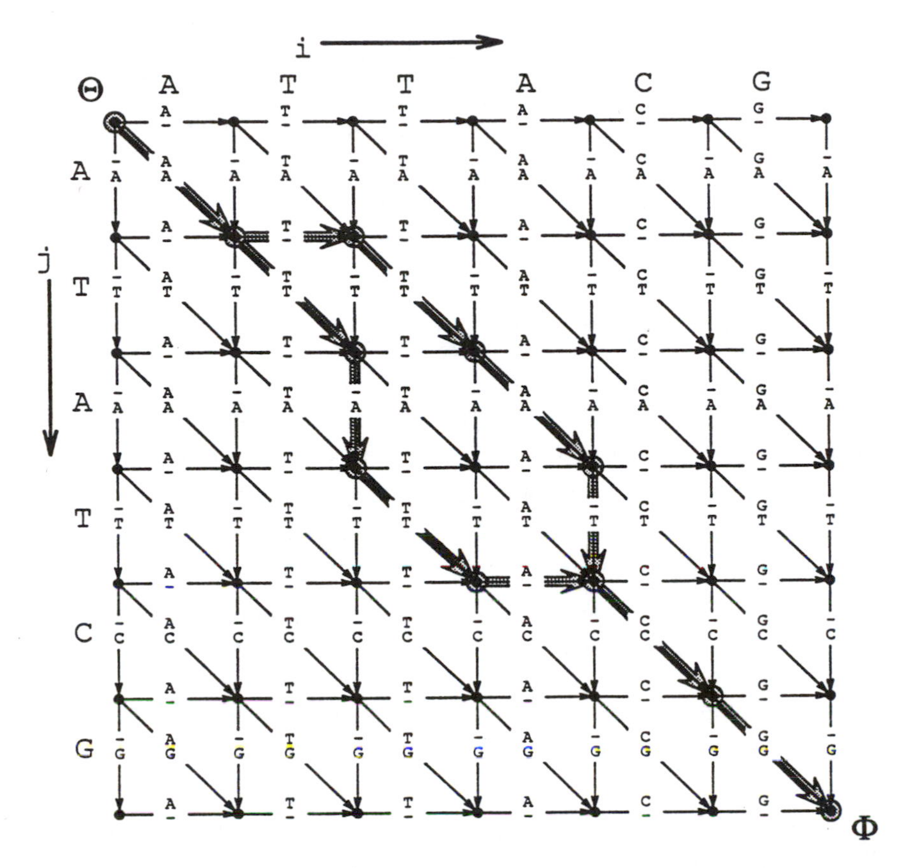

FIGURE 3.2 $G_{\mathbf{A,B}}$ for **A** = ATTACG and **B** = ATATCG.

subsection. First, observe that, in terms of the edit graph formulation, we seek the score of a maximal-score path from the vertex Θ at the upper left-hand corner of the graph $G_{\mathbf{A,B}}$ to the vertex Φ at the lower right-hand corner.

Consider computing $S(i,j)$, the score of a maximal-score path from Θ to some given vertex (i,j) in the graph. Because there are only three edges directed into vertex (i,j), it follows that any optimal path P to (i,j) must fit one of the following three cases: (1) P is an optimal path to $(i-1,j)$ followed by the **A**-gap edge into (i,j); (2) P is an optimal path to $(i,j-1)$ followed by the **B**-gap edge into (i,j); or (3) P is an optimal path to

$(i-1, j-1)$ followed by the alignment edge into (i, j). It is critical to note that the subpath preceding the last edge of *P* must also be optimal, for, if it is not, then it is easy to show that *P* cannot be optimal, a contradiction. This observation immediately leads to the fundamental recurrence:

$$\begin{aligned} S(i, j) = \max \{ & S(i-1, j-1) + \delta(a_i, b_j), \\ & S(i-1, j) + \delta(a_i, -), \\ & S(i, j-1) + \delta(-, b_j)\}, \end{aligned}$$

which states that the maximal score of a path to (i, j) is the larger of (1) the maximal score of a path to $(i-1, j)$ plus the score of the **A**-gap edge to (i, j), (2) the maximal score of a path to $(i, j-1)$ plus the score of the **B**-gap edge to (i, j), or (3) the maximal score of a path to $(i-1, j-1)$ plus the score of the alignment edge to (i, j).

All that is needed to have an effective computational procedure based on this recurrence is to determine an order in which to compute *S*–values. There are many possible orders. Three simple alternatives are (1) column by column from left to right, top to bottom in each column, (2) row by row from top to bottom, left to right in each row, and (3) antidiagonal by antidiagonal from the upper left to the lower right, in any order within an antidiagonal (antidiagonal k consists of the vertices (i, j) such that ($i + j = k$). Using the first sample ordering leads to the algorithm of Figure 3.3. In this algorithm, *M* denotes the length of **A** and *N* denotes the length of **B**.

The algorithm of Figure 3.3 computes $S(i, j)$ for every vertex (i, j) in an $(M+1) \times (N+1)$ matrix in the indicated order of *i* and *j*. Along the left and upper boundaries of the edit graph (that is, vertices with $i = 0$ or $j = 0$, respectively), the algorithm utilizes the recurrence, except that terms referencing nonexistent vertices are omitted (that is, in lines 3 and 5, respectively). The algorithm of Figure 3.3 takes $O(MN)$ time; that is, when M and N are sufficiently large, the time taken by the algorithm does not grow faster than the quantity *MN*. If one stores the whole

```
0.  var S: array [0..M,0..N] of real
1.  S[0,0] ← 0
2.  for j ← 1 to N do
3.        S[0, j] ← S[0, j − 1] + δ(−, b_j)
4.  for i ← 1 to M do
5     { S[i,0] ← S[i − 1,0] + δ(a_i,−)
6.        for j ← 1 to N do
7.              S[i, j] ← max   {S[i − 1, j − 1] + δ(a_i, b_j)
                                  S[i − 1, j] + δ(a_i,−),
                                  S[i, j − 1] + δ(−, b_j)}
8.    }
9.  write "Maximum score is" S[M, N].
```

FIGURE 3.3 The classical dynamic programming algorithm.

$(M+1)\times(N+1)$ matrix S, then the algorithm also requires $O(MN)$ space.

The algorithm of Figure 3.3 is a *dynamic programming* algorithm that utilizes the fundamental recurrence. Dynamic programming is a general computational paradigm of wide applicability (see, for example, Horowitz and Sahni, 1978). A problem can be solved by dynamic programming if the final answer can be efficiently determined by computing a tableau of optimal answers to progressively larger and larger subproblems. The *principle of optimality* requires that the optimal answer to a given subproblem be expressible in terms of optimal answers to smaller subproblems. Our basic sequence comparison problem does yield to this principle: the optimal answer $S(i,j)$ for the problem of comparing prefix $A_i = a_1a_2\ldots a_i$ and prefix $B_j = b_1b_2\ldots b_j$ can be found by computing optimal answers for smaller prefixes of **A** and **B**. The recurrence formula describes the relationship of each subproblem to a larger subproblem.

The algorithm of Figure 3.3 computes only the score of a maximum-scoring global alignment between **A** and **B**. One or all of these optimal alignments can be recovered by tracing the paths backwards from

Φ to Θ with the aid of the now complete matrix *S*. Specifically, an edge from vertex v_1 to Φ is on an optimal path if $S(v_1)$ plus the score of its edge equals $S(\Phi)$. If v_1 is on an optimal path, then, in turn, an edge from v_2 to v_1 is on an optimal path if $S(v_2)$ plus the score of the edge equals $S(v_1)$. In this way, one can follow an optimal path back to the start vertex Θ. In essence, this *traceback* procedure moves backwards from a vertex to the preceding vertex whose term in the three-way maximum of the recurrence yielded the maximum. The possibility of ties creates the possibility of more than a single optimal path. Unfortunately, this traceback technique for identifying one or more optimal paths requires that the entire matrix *S* be retained, giving an algorithm that takes $O(MN)$ space as well as time.

A more space-efficient approach to delivering an optimal alignment begins with the observation that if only the score of an optimal alignment is desired then only the value of $S(M,N)$ is needed, and so *S*-values can be discarded once they have been used in computing the values that depend on them. Observing that one need only know the previous column in order to compute the next one, it follows that only two columns need be retained at any instance, and so only $O(N)$ space is required. Such a *score-only* algorithm can be used as a subprocedure in a divide-and-conquer algorithm that determines *an* optimal alignment using only $O(M+N)$ space. The divide step consists of finding the midpoint of an optimal source-to-sink path by running the score-only algorithm on the first half of **B** and the reverse of the second half of **B**. The conquer step consists of determining the two halves of this path by recursively reapplying the divide step to the two halves. Myers and Miller (1988) have shown this strategy to apply to most comparison algorithms that have linear-space score-only algorithms. This refinement is very important, since space, not time, is often the limiting factor in computing optimal alignments between large sequences. For example, two sequences of length 100,000 can be compared in several hours of CPU time, but would require 10 billion units of memory if optimal alignments were delivered using the simple $O(MN)$ space traceback approach. This is well beyond the memory capacity of any conventional machine.

FINDING LOCAL SIMILARITIES

We now turn to the problem of finding local alignments, that is, subsegments of **A** and **B** that align with maximal score. Local alignments can be visualized as paths in the edit graph, $G_{\mathbf{A},\mathbf{B}}$. Unlike the global alignment problem, the path may start and end at any vertices, not just from Θ and Φ. Intrinsic to determining local similarities is the requirement that the scoring scheme δ be designed with a negative bias. That is, for alignment of unrelated sequences (under some suitable stochastic model of the sequences) the score of a path must on the average be negative. If this were not the case, then longer paths would tend to have higher scores, and one would generally end up reporting a global alignment between two sequences as the optimal local alignment. For example, the simple scoring scheme of Figure 3.1 is not negatively biased, whereas the scheme of Figure 3.4 is. Note that under this new scheme, the alignment $\begin{matrix}\texttt{ATTACG}\\ \texttt{ATATCG}\end{matrix}$ is now optimal with score 3.34, whereas $\begin{matrix}\texttt{ATTA-CG}\\ \texttt{A-TATCG}\end{matrix}$ now has lesser score 3. In this case, the optimal alignment happened to be global, but for longer sequences this is generally not the case. For example, the best local alignment between GAGG<u>TTGCTGA</u>GAA and ACTCTTC<u>TTCCTTA</u> is the alignment $\begin{matrix}\texttt{TTGCTGA}\\ \texttt{TTCCTTA}\end{matrix}$ of score 4.34 between the underlined substrings.

δ	–	A	C	G	T
–	-1	-1	-1	-1	-1
A	-1	1	-.33	-.33	-.33
C	-1	-.33	1	-.33	-.33
G	-1	-.33	-.33	1	-.33
T	-1	-.33	-.33	-.33	1

FIGURE 3.4 A local-alignments scoring scheme.

The design of scoring schemes that properly weigh alignments to expose biologically meaningful local similarities is the subject of much investigation. The score of alignments between protein sequences is the sum of scores assigned to individual pairs of aligned symbols, just as for

DNA. However, since proteins are represented by combinations of 20 letters and the gap symbol, the table of scores is now 21×21. These scores may be chosen by users to fit the notion of similarity they have in mind for the comparison. For example, Dayhoff et al. (1983) compiled statistics on the frequency with which one amino acid would mutate into another over a fixed period of time and from these built a table of aligned symbol scores consisting of the logarithm of the normalized frequencies. Under Dayhoff's scoring scheme, the score of an alignment is a coarse estimate of the likelihood that one segment has mutated into the other. Figure 3.5 is a scaled integer approximation of Dayhoff's matrix that is much used in practice today.

The basic issue in local alignment, just as in the case of global alignment, is to find a path of maximal score. However, there are more degrees of freedom in the local alignment problem: where the paths begin and where they end is not given a priori but is part of the problem. Note that if we knew the vertex (g,h) at which the best path began, we could find its score and end-vertex by setting $S(g,h)$ to 0 and then applying the fundamental recurrence to all vertices (i,j) for which $i \geq g$ and $j \geq h$. We can capture *all* potential start vertices simultaneously by modifying the central recurrence so that 0 is a term in the computation of the maximum; that is,

$$
\begin{aligned}
S(i,j) = \{0, & S(i-1,j-1) + \delta(a_i, b_j), \\
& S(i-1,j) + \delta(a_i, -), \\
& S(i,j-1) + \delta(-, b_j)\}.
\end{aligned}
$$

Indeed, with this simple modification, $S(i,j)$ is now the score of the highest-scoring path to (i,j) that begins at *some* vertex (g,h) for which $g \leq i$ and $h \leq j$. The best score of a path in the edit graph is then the maximum over all vertices in the graph of their *S*-values. A vertex achieving this maximum is the end of an optimal path. This basic result is often referred to as the Smith-Waterman algorithm after its inventors (Smith and Waterman, 1981). The beginning of the path, the segments it aligns, and the alignment between these segments can all be delivered in linear space by further extensions of the treatment given above for global alignments. If one uses such a comparison algorithm with the scoring

δ	–	A	R	N	D	C	Q	E	G	H	I	L	K	M	F	P	S	T	W	Y	V
–	–8	–8	–8	–8	–8	–8	–8	–8	–8	–8	–8	–8	–8	–8	–8	–8	–8	–8	–8	–8	–8
A	–8	3	–3	0	0	–3	–1	0	1	–3	–1	–3	–2	–2	–4	1	1	1	–7	–4	0
R	–8	–3	6	–1	–3	–4	1	–3	–4	1	–2	–4	2	–1	–4	–1	–1	–2	1	–6	–3
N	–8	0	–1	4	2	–5	0	1	0	2	–2	–4	1	–3	–4	–2	1	0	–5	–2	–3
D	–8	0	–3	2	5	–7	1	3	0	0	–3	–5	–1	–4	–7	–2	0	–1	–8	–5	–3
C	–8	–3	–4	–5	–7	9	–7	–7	–5	–4	–3	–7	–7	–6	–6	–3	–1	–3	–8	–1	–2
Q	–8	–1	1	0	1	–7	6	2	–3	3	–3	–2	0	–1	–6	0	–2	–2	–6	–5	–3
E	–8	0	–3	1	3	–7	2	5	–1	–1	–3	–4	–1	–4	–6	–1	–1	–2	–8	–4	–3
G	–8	1	–4	0	0	–5	–3	–1	5	–4	–4	–5	–3	–4	–5	–2	1	–1	–8	–6	–2
H	–8	–3	1	2	0	–4	3	–1	–4	7	–4	–3	–2	–4	–2	–1	–2	–3	–5	–1	–3
I	–8	–1	–2	–2	–3	–3	–3	–3	–4	–4	6	1	–2	1	0	–3	–2	0	–7	–2	3
L	–8	–3	–4	–4	–5	–7	–2	–4	–5	–3	1	5	–4	3	0	–3	–4	–3	–5	–3	1
K	–8	–2	2	1	–1	–7	0	–1	–3	–2	–2	–4	5	0	–6	–2	–1	–1	–5	–6	–4
M	–8	–2	–1	–3	–4	–6	–1	–4	–4	–4	1	3	0	8	–1	–3	–2	–1	–7	–4	1
F	–8	–4	–4	–4	–7	–6	–6	–6	–5	–2	0	0	–6	–1	8	–5	–3	–4	–1	4	–3
P	–8	1	–1	–2	–2	–3	0	–1	–2	–1	–3	–3	–2	–3	–5	6	1	–1	–7	–6	–2
S	–8	1	–1	1	0	–1	–2	–1	1	–2	–2	–4	–1	–2	–3	1	3	2	–2	–3	–2
T	–8	1	–2	0	–1	–3	–2	–2	–1	–3	0	–3	–1	–1	–4	–1	2	4	–6	–3	0
W	–8	–7	1	–5	–8	–8	–6	–8	–8	–5	–7	–5	–5	–7	–1	–7	–2	–6	12	–1	–8
Y	–8	–4	–6	–2	–5	–1	–5	–4	–6	–1	–2	–3	–6	–4	4	–6	–3	–3	–1	8	–3
V	–8	0	–3	–3	–3	–2	–3	–3	–2	–3	3	1	–4	1	–3	–2	–2	0	–8	–3	5

FIGURE 3.5 A protein local-alignment scoring scheme.

scheme of Figure 3.5, one sees the three regions of similarity shown in Figure 3.6 between the sequence of the monkey somatotropin protein and the somatotropin precursor protein of a rainbow trout. Note that while in many cases the aligned symbols are identical, they do not have to be.

Thus far we have presented the local similarity problem as one of finding two subsegments of the sequences that align with maximal score. But as illustrated in Figure 3.6, the ultimate goal is to expose not a single such alignment, but all the significantly conserved segments, ideally nonoverlapping as in the somatotropin example of Figure 3.5. To this end, Waterman and Eggert (1987) proposed the following simple algorithm. Find a highest-scoring local alignment by the method indicated in the previous paragraph. Eliminate every edge in the edit graph that is adjacent to a vertex on the path of this local alignment. Now find a highest-scoring path over the remaining graph. Eliminate the edges adjacent to this second-best path, and proceed to find a third-best path, and so on. In this way, one produces a series of local alignments of decreasing score whose underlying paths do not intersect. Note that this procedure may generate local alignments whose substrings overlap. Nonetheless, this procedure is very effective in identifying the biologically relevant local homologies between two sequences. As originally presented, the algorithm requires $O(MN)$ space, but recent refinements by Chao and Miller (1994) have reduced both storage and computing time and have permitted the comparison of two sequences of length 100,000 on a conventional workstation in several hours.

The output of such a problem could be displayed as a sequence of alignments, as in Figure 3.6. It is also convenient and illuminating to depict all the alignments as paths in an edit graph, as in Figure 3.2. However, as the sequences become larger and larger, one must "step back" from the details of the edit graph. Figure 3.7 is a depiction of the edit graph of the monkey and rainbow trout somatotropin sequences of Figure 3.6 where only the paths corresponding to the three aligned segment pairs are drawn. At this level of resolution, the small gaps in the alignments of the second and third segment pairs appear as small discontinuities in paths that otherwise follow the direction of the diagonal of the edit graph grid or matrix. When the sequences become very large, say on the order of 100,000 nucleotides, then small local alignments are not seen, and neither are gaps in large alignments unless they are very large. Nonetheless, such

```
Score = 69

Position:  2         *         *         *         *41
Monkey:    PTIPLSRLFDNAMLRAHRLHQLAFDTYQEFEEAYIPKEQK
Trout:     SAIENQRLFNIAVSRVQHLHLLAQKMFNDFDGTLLPDERR
Position: 10          *         *         *        49

Score = 103

Position:  51        *         *         *         *         *         *         *         *131
Monkey:     SLCFSESIPTPSNREETQQKSNLQLLRISLLLIQSWLEPVQFLRSVFANSLVYGTSYSDVYDLLKDLEEGIQTLMGRLEDG
Trout:      DFCNSDSIVSPVDKHETQKSSVLKLLHISFRLIESWEYPSQTL--IISNSLMVRNA-NQISEKLSDLKVGINLLITGSQDG
Position:  58 *         *         *         *         *           *          *         *   135

Score = 115

Position: 143       *         *         *         *        189
Monkey:     YSKFDTNSHNDDALLKNYGLLYCFRKDMDKIETFLRIVQCR-SVEGSC
Trout:      YGNYYQNLGGDGNVRRNYELLACFKKDMHKVETYLTVAKCRKSLEANC
Position: 150          *         *         *         *     197
```

FIGURE 3.6 Conserved regions of two somatotropin proteins.

dot plots give a meaningful visualization of all the similarities between segments in a single snapshot and are ubiquitous.

VARIATIONS ON SEQUENCE COMPARISON

In this section a number of the most important variations on sequence comparison are examined. The survey is by no means exhaustive.

Variations in Gap Cost Penalties

How to assign scores to alignment gaps has always been more problematic than scoring aligned symbols, because the statistical effect of gaps is not well understood (see Chapter 4). Nature frequently deletes or inserts entire substrings as a unit, as opposed to individual polymer

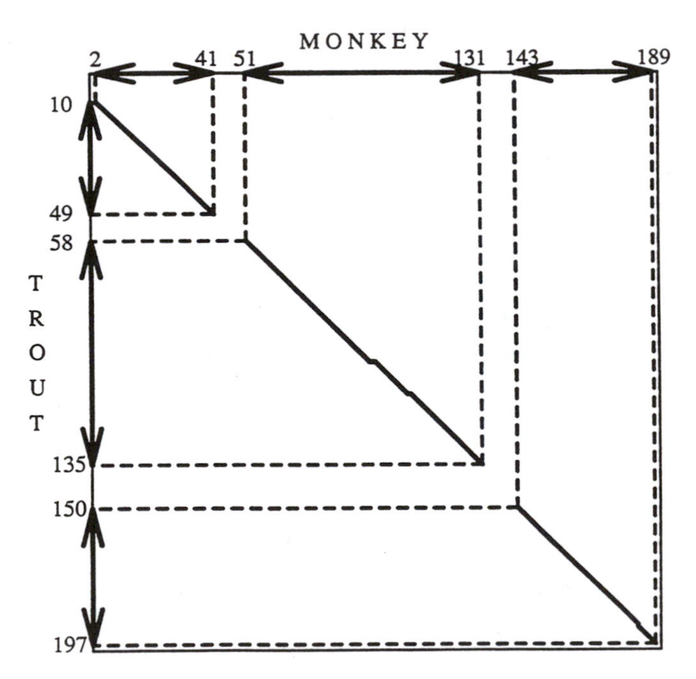

FIGURE 3.7 Dot plot of somatotropin alignments.

elements. It is thus natural to think of cost models in which the score of a gap is not just the sum of scores assigned to the individual symbols in the gap, as was used in the previous two sections, but rather a more general function, *gap(x)*, of its length x. For example, it is common to score a gap according to the *affine* function $gap(x) = r + sx$, where $r > 0$ is the penalty for the introduction of the gap and $s > 0$ is the penalty for each symbol in the gap. Such affine gap costs are particularly important when comparing proteins. For example, a gap penalty of $8 + 4x$ works well in conjunction with the aligned symbol scores of Figure 3.5. Because a gap is viewed as detracting from similarity, its score is a penalty that is *subtracted* from the total.

Accommodating affine gap scores involves the following variation on the central recurrence (Gotoh, 1982). For each subproblem, $\mathbf{A}_i$ versus $\mathbf{B}_j$, one develops recurrences for (1) the best alignment *that ends with an A-gap,* $Ag(i,j)$, (2) the best alignment *that ends with a B-gap,* $Bg(i,j)$, and (3) the best overall alignment, $S(i,j)$. This leads to the following system of recurrence equations:

$$
\begin{aligned}
Ag(i,j) &= \max\{Ag(i-1,j)-s, S(i-1,j)-(r+s)\} \\
Bg(i,j) &= \max\{Bg(i,j-1)-s, S(i,j-1)-(r+s)\} \\
S(i,j) &= \max\{S(i-1,j-1)+\delta(a_i,b_j), Ag(i,j), Bg(i,j)\}.
\end{aligned}
$$

S terms contributing to an *Ag* or *Bg* value are penalized $r + s$ because a gap is being initiated from that term. *Ag* terms contributing to *Ag* values and *Bg* terms contributing to *Bg* values are penalized only *s* because the gap is just being extended. An algorithm that applies these recurrences at each (i,j) leads to an $O(MN)$ time algorithm for global alignments with affine gap costs. Simply adding a 0 term to the *S*-recurrence gives an algorithm for local alignments with affine gap costs.

Summation and affine functions are not the only options available for scoring gaps. The gap cost function *gap*(*x*) can be taken to be a concave (flat or cupped downward) function of length, that is, a function such that $gap(x+1) - gap(x) \leq gap(x) - gap(x-1)$ for all $x > 0$. The class of concave gap cost functions includes affine functions but is much wider than just affine functions. For example, for positive *a* and *b*, the function $gap(x) = a \log x + b$ is a concave function that finds occasional use in sequence comparison. It has been postulated that such a model is natural for biological sequences where gap costs would be expected to have a decreasing marginal penalty as a function of length. For this model, investigators have been able to design algorithms that take $O(MN(\log N + \log M))$ time or less (Miller and Myers, 1988; Eppstein et al., 1989).

It is also possible to design an algorithm for completely arbitrary gap cost functions. However, such generality comes at a price: the best available algorithm takes $O(MN(M+N))$ time (Waterman et al., 1976). For this reason and because the more restricted affine and concave models appear adequate to most needs, the general algorithm is rarely used.

The Duality Between Similarity and Difference Measures

Thus far we have considered the comparison problem to be one of exposing the similarity between two sequences and thus have naturally thought in terms of *maximizing* the score of alignments. Another natural perspective is to think about how a sequence **A** may have evolved into sequence **B** over time. In this context, one seeks alignments that reveal the *minimum* number of mutational events that might have effected the transformation. In this view, an aligned symbol of **B** is *substituted* for its counterpart in **A**, an unaligned symbol in **A** is *deleted*, and an unaligned symbol in **B** is *inserted*. For example, in the alignment $\begin{smallmatrix}\texttt{ATTA-CG}\\\texttt{A-TATCG}\end{smallmatrix}$, the first `T` in `ATTACG` is deleted, and the second `T` in `ATATCG` is inserted. In the alignment $\begin{smallmatrix}\texttt{ATTACG}\\\texttt{ATATCG}\end{smallmatrix}$, `T` is mutated into `A`, and `A` is mutated into `T`. As before, the scoring scheme δ assigns a score to each evolutionary event modeled by a column, but now the interpretation is that δ represents the differences rather than the similarities between symbols. Note that for formal purposes it is assumed that an `A` mutates into an `A` in the alignments above at no cost; that is, one chooses $\delta(\texttt{A},\texttt{A})$ to be 0.

Given a scoring scheme δ reflecting an evolutionary or difference-based model, the goal is to find an alignment of minimal score, that is, one that indicates the minimum-scoring set of changes needed to go from one sequence to a related sequence. Let $D(\mathbf{A},\mathbf{B})$ be the score of a minimal cost alignment between sequences **A** and **B**. In honor of its inventor, this score is formally known as the *generalized Levenshtein measure* or *distance* between sequences **A** and **B**. Indeed the measure, *D*, between sequences forms a metric space over sequences if the underlying scoring function δ forms a metric space over the underlying alphabet. Thus calling this measure a distance is formally correct for a wide class of scoring schemes δ.

Immediately note that the distance and similarity perspectives are complementary. To solve a difference problem, we need only revise our previous discussions by replacing *maximum* with *minimum* in every sentence and formula. Also, one could simply take a δ for a difference problem and multiply every score by -1. Applying the similarity algorithm with the modified scores would produce optimal alignments for the original difference problem, and multiplying the resultant similarity score by -1 would give the distance between the two sequences.

Aligning More Than Two Sequences at a Time

Molecular biologists are frequently interested in comparing more than two sequences simultaneously. For instance, given a number of sequences of the same functionality, it is much more likely that the similarity that gives this common function will be more evident among the group than among two sequences from the group. A closely related problem is to discover the evolutionary relationships between a set of sequences by constructing an evolutionary tree, or *phylogeny*, that minimizes the evolutionary changes that must have taken place along each branch of the tree. A third application for aligning a collection of sequences is to correct errors in the "raw" experimental data obtained in DNA sequencing experiments. Typically, 1 to 10 percent of the symbols in a sequenced fragment are incorrect, missing, or spurious. These errors are detected and corrected by sequencing a given stretch several times and then forming a consensus by aligning the sequences. Figure 3.8 illustrates a multi-alignment of such sequence data.

```
CTCGCG-CACAT-AGGGCG-GTC-CGAGA-GA-TAGGCAAGCC
CTCGCGGCACATTCGGGCG-GTCTCGAGATGACTAGGC-AGCC
CTCGCGGCA-ATTCGGGCG-GTCTCGAGA-GACTAGGCAAGCC
CTCGCGTCACATTCGGGCGTGTCTCGAGA-GACTAGGCAAGCC
CTCGCGG-ACATTCGGGCG-GT-TCG-GA-GACTAGGCAAGCC
-------------------------------------------

CTCGCGGCACATTCGGGCG GTCTCGAGA GACTAGGCAAGCC←"consensus"
```

FIGURE 3.8 A multi-alignment of five DNA sequences.

Suppose we wish to align K sequences $\mathbf{A}^1, \mathbf{A}^2, \ldots, \mathbf{A}^K$, where $\mathbf{A}^i = a_1^i a_2^i \cdots a_{N^i}^i$ is of length N^i. As for the basic problem, we wish to arrange the sequences into a tableau using dashes to force the alignment of certain characters in given columns. For example, in Figure 3.8 the dashes are placed so as to arrange columns consisting of primarily one symbol. For each column, the consensus of the column is the symbol that occurs the greatest number of times in that column. Concatenating these consensus characters together, ignoring dashes, gives the consensus sequence for the five experimental trials. As for pairwise alignments, each column of K symbols of the multi-alignment is scored according to a user-supplied function δ. For example, if δ is the number of symbols in the column not equal to the majority symbol of the column (which can be a dash), then the multi-alignment of Figure 3.8 has score 13, and this is the minimum possible score over all possible multi-alignments of the five sequences.

The problem of finding a maximum (minimum)-scoring alignment among K sequences can be solved by extending the dynamic programming recurrence for the basic problem from a recurrence over a two-dimensional matrix to a recurrence over a K-dimensional matrix. Let $\mathbf{i} = (i_1, i_2, \ldots, i_K)$ be a vector in K-dimensional Cartesian space. Now we compute a K-dimensional array S, where $S(\mathbf{i})$ is the score of the best alignment among the prefix sequences $\mathbf{A}_{i_1}^1, \mathbf{A}_{i_2}^2, \ldots, \mathbf{A}_{i_K}^K$. The central recurrence now becomes

$$S(\mathbf{i}) = \max\{S(\mathbf{i} - \mathbf{e}) + \delta(e_1 : a_{i_1}^1, \ldots, e_K : a_{\mathbf{i}_K}^K) : \mathbf{e} \in \{0,1\}^K - (0,0,\ldots,0)\},$$

where $e{:}a$ means "if $e = 1$ then a else '–' ".

In terms of an edit graph model, imagine a grid of vertices in K-dimensional space where each vertex **i** has $2^K - 1$ edges directed into it, each corresponding to a column that when appended to the alignment for the edge's tail gives rise to the alignment for the prefix sequences represented by **i**. Computing the S values in some topological ordering requires a total of $O(N^K)$ time, where $N = \max_i N^i$. While multiple sequence comparison algorithms of this genre are conceptually straightforward, they take an exponential amount of time in K and are thus generally impractical for $K > 3$.

Multiple sequence comparison has been shown to be NP-complete (Garey and Johnson, 1979), which means that it is almost surely the case that any algorithm for this problem must exhibit time behavior that is exponential in K. Thus many authors have sought heuristic approximations, the most popular of which is to take $O(K^2 N^2)$ time to compute all pairwise optimal alignments between the K sequences, and then produce a multiple sequence alignment by merging these pairwise alignments. Note that any multiple sequence alignment induces an alignment between a given pair of sequences (take the two rows of the tableau and remove any columns consisting of just dashes). However, given all of the possible $K(K-1)/2$ pairwise alignments between K sequences, it is almost always impossible to arrange a multi-alignment consistent with them all. Try, for example, merging the best pairwise alignments among ACG, CGA, and GAC. But, given any $K-1$ alignments relating all the sequences (that is, a spanning tree of the complete graph of sequence pairs), it is always possible to do so. Feng and Doolittle (1987) compare a number of methods based on this approach. The most recent algorithms utilize the natural choice of the $K-1$ alignments whose scores sum to the minimal possible amount (that is, a minimum spanning tree of the complete graph of sequence pairs). However, such merges do not always lead to optimal alignments, as is illustrated by the following example:

```
G-CACA                   G--CACA      G-CACA
GGCA-A and GG-CAA yield  GG-CA-A, but GGCA-A is better.
           GGACA-        GGACA--      GG-ACA
```

While the choice of δ for a multi-alignment scoring scheme is conceptually a function of K arguments, it is often the case that δ is effectively defined in terms of an underlying pairwise scoring function δ'. For example, the *sum-of-pairs score* is defined as $\delta(a_1, a_2, \ldots, a_K) = \sum_{i<j} \delta'(a_i, a_j)$, where one must let $\delta(-,-) = 0$. In essence, the sum-of-pairs multi-alignment score is the sum of the scores of the $K(K-1)/2$ pairwise alignments it induces. Another common scheme is the *consensus score,* which defines $\delta(a_1, a_2, \ldots, a_K)$ as max/min$\{\sum_i \delta'(c, a_i) : c \in \Psi \cup \{-\}\}$. The symbol c that gives the best score is said to be the consensus symbol for the column, and the concatenation of these symbols is the consensus sequence. In effect, the consensus multi-alignment score is the sum of the scores of the K pairwise alignments of the sequences versus the consensus. The example of Figure 3.8 is such a scoring scheme where δ' is the scoring scheme of Figure 3.1. While we do not show it here, the problem of determining minimal phylogenies mentioned at the start of this subsection can also be modeled as an instance of a multiple sequence alignment problem by choosing a δ for columns that suitably encodes the tree relating the sequences (Sankoff, 1975). However, the more general phylogeny problem requires that one also determine the tree that produces the minimal score. This daunting task essentially requires the exploration of the space of all possible trees with K vertices. So in practice, evolutionary biologists have put a great deal of effort into designing heuristic algorithms for the phylogeny problem, and there is much debate about which of these is best.

K-Best Alignments

The alignment algorithm in the section "The Basic Dynamic Programming Algorithm" above reports an optimal alignment that is clearly a function of the choice of scoring scheme. Unfortunately, biologists have not yet ascertained which scoring schemes are "correct" for a given comparison domain. This uncertainty has suggested the problem of listing all alignments near the optimum in the hope of generating the biologically correct alignment.

From the point of view of the edit graph formulation, the K-best problem is to deliver the K-best shortest source-to-sink paths, a problem much studied in the operations research literature. Indeed, there is an $O(MN+KN)$ time and space algorithm, immediately available from this literature (Fox, 1973), that delivers the K-best paths over an edit graph. The algorithm delivers these paths/alignments in order of score, and K does not need to be known a priori: the next best alignment is available in $O(N)$ time. The essential idea of the algorithm is to keep, at each vertex v, an ordered list of the score of the next best path to the sink through each edge out of v. The next best path is traceable using these ordered lists and is extracted, and the lists are appropriately updated.

If all one desires is an enumeration, not necessarily in order of score, of all alignments that are within ε of the optimal difference $D(\mathbf{A},\mathbf{B})$, then a simpler method is available that requires only the matrix S of the dynamic programming computation. While not any faster in time, the simpler alternative below does require only $O(MN)$ space. One can imagine tracing back all paths from the sink to the source in a recursive fashion. The essential idea of the algorithm is to limit the traceback to only those paths of score not greater than $D(\mathbf{A},\mathbf{B})+\varepsilon$. Suppose one reaches vertex (i,j) and the score of the path thus far traversed from the sink to this vertex is $T(i,j)$. Then one traces back to predecessor vertices $(i-1,j)$, $(i-1,j-1)$, and $(i,j-1)$ if and only if:

$$
\begin{aligned}
&S(i-1,j)+\delta(a_i,-)+T(i,j)\le D(\mathbf{A},\mathbf{B})+\varepsilon,\\
&S(i-1,j-1)+\delta(a_i,b_j)+T(i,j)\le D(\mathbf{A},\mathbf{B})+\varepsilon,\\
&S(i,j-1)+\delta(-,b_j)+T(i,j)\le D(\mathbf{A},\mathbf{B})+\varepsilon,
\end{aligned}
$$

respectively. This procedure is very simple, space economical, and quite fast.

A classic example of the need for affine gap costs was presented in a paper by Smith and Fitch (1983) comparing the α and β chicken hemoglobin chains. For a setting of the gap costs that gave the biologically correct alignment, there were 17 optimal alignments, 1,317 alignments within 5 percent of the optimum, and 20,137,655 within 20 percent of the optimum. This kind of exponential growth suggests that perhaps rather than list alignments, one should report the best possible scores in order or give a color-coded visualization of the edit graph that colors edges according to the score of the best path utilizing the edge. Another interesting variation is to explore the range of solutions not by enumerating near-optimal answers, but by studying the range of optimal answers produced by parametrically varying aspects of the underlying scoring scheme (Waterman et al., 1992).

Approximate Pattern Matching

A variation on the local alignments problem discussed above is the approximate match problem. For this problem, imagine that **A** is a very long sequence and **B** a comparatively short query sequence. The problem is to find substrings of **A**, called match sites, that align with the entirety of **B** with a score greater than some user-specified threshold. An example might be to find all locations in a chromosome's DNA sequence (**A**) where a particular DNA sequence element (**B**) or some sequence like it occurs. It is not hard to see that this problem is equivalent to finding sufficiently high scoring paths that begin at a vertex in row 0 and end at row N of the edit graph for **A** and **B**. By simply permitting 0 to be a term in the computation of S-values in row 0 and checking values in row N, one obtains the desired modification of the basic dynamic programming algorithm.

The problem is taken to another level by generalizing **B**, the query, from a sequence to a pattern (that describes a set of sequences). This variation is called approximate pattern matching. Computer scientists working on text-searching applications have long studied the problem of finding exact matches to a pattern in a long text. That is, given a pattern as a query, and a text as a database, one seeks substrings of the database text that match the pattern (exactly). Pattern types that have been much studied include the cases of a simple sequence, a regular expression, and a context-free language. Such patterns are notations that denote a possibly infinite set of sequences, each of which is said to (exactly) match the pattern. For example, the regular expression `A(T|C)G*` denotes the set of sequences that start with an `A` followed by a `T` or a `C` and then zero or more `G`'s, that is, the set {`AT`, `AC`, `ATG`, `ACG`, `ATGG`, `ACGG`, `ATGGG`,...}. Assuming the pattern takes P symbols to specify and the text is of length

N, there are algorithms that solve the text searching problem in $O(P+N)$, $O(PN)$, and $O(PN^3)$ time, depending on whether the pattern is a simple sequence, a regular expression, or context-free language, respectively. Fusing the concept of exact pattern matching and sequence comparison gives rise to the class of approximate pattern matching problems. Given a pattern, a database, a scoring scheme, and a threshold, one seeks all substrings of the database that align to some sequence denoted by the pattern with score better than the threshold. In essence, one is looking for substrings that are within a given similarity neighborhood of an exact match to the pattern. Within this framework, the similarity search problem is an approximate pattern matching problem where the pattern is a simple sequence. We showed earlier that this problem can be solved in $O(PN)$ time. For the case of regular expressions, the approximate match problem can also be solved in $O(PN)$ time (Myers and Miller, 1989), and, for context-free languages, an $O(PN^3)$ algorithm is known (Myers, 1994a). While the cost of searching for approximate matches to context-free languages is prohibitive, searching for approximate matches to regular expressions is well within reason and finds applications in searching for matches to structural patterns that occur in proteins.

Parallel Computing

The basic problem of comparing sequences has resisted better than quadratic, $O(MN)$ time algorithms. This has led several investigators to study the use of parallel computers to achieve greater efficiency. As stated above, the *S*-matrix can be computed in any order consistent with the data dependencies of the fundamental recurrence. One naturally thinks of a row-by-row or column-by-column evaluation, but we pointed out as a third alternative that one could proceed in order of antidiagonals. Let antidiagonal *k* be the set of entries $\{(i,j)\colon\ i+j=k\}$. Note that to compute antidiagonal *k*, one only needs antidiagonals $k-1$ and $k-2$. The critical observation for parallel processing is that each entry in this antidiagonal can be computed independently of the other entries in the antidiagonal, a fact not true of the row-by-row and column-by-column evaluation procedures. For large SIMD (single-instruction, multiple-data) machines, a processor can be assigned to each entry in a fixed antidiagonal and compute its result independently of the others. With $O(M)$ processors, each antidiagonal can be computed in constant time, for a total of $O(N)$ total elapsed time. Note that total work, which is the product of processors and time per processor, is still $O(MN)$. The improvement in time stems from the use of more processors, not from an intrinsically more efficient algorithm.

This observation about antidiagonals has been used to design custom VLSI (very large scale integration) chips configured in what is called a *systolic array*. The "array" consists of a vector of processors, each of which is identical, performs a dedicated computation, and communicates only with its left and right neighbors, making it easy to lay out physically on a silicon wafer. For sequence comparisons, processor i computes the entries for row i and contains three registers that we will call $L(i)$, $V(i)$, and $U(i)$. At the completion of the k th step, the processors contain antidiagonals k and $k-1$ in their L and V registers, respectively, and the characters of **B** flow through their U registers. That is, $L(i)_k = S(i, k-i-1)$, $V(i)_k = S(i, k-i)$, and $U(i)_k = b_{k-i}$, where $X(i)_k$ denotes the value of register X at the end of the k th step. It follows from the basic recurrence for S-values that the following recurrences correctly express the values of the registers at the end of step $k+1$ in terms of their values at the end of step k:

$$
\begin{aligned}
U(i)_{k+1} &= U(i-1)_k, \\
L(i)_{k+1} &= V(i)_k, \\
V(i)_{k+1} &= \max\{L(i-1)_k + \delta(a_i, U(i-1)_k), \\
&\qquad V(i-1)_k + \delta(a_i, -), \\
&\qquad V(i)_k + \delta(-, U(i-1)_k)\}.
\end{aligned}
$$

These recurrences reveal that to accomplish step $k+1$, processor $i-1$ must pass its register values to processor i and each processor must have just enough hardware to perform three additions and a three-term maximum. Moreover, each processor must have a $(2|\Psi|+1)$-element memory that can be loaded with the scores for $\delta(a_i,?)$, $\delta(-,?)$, and $\delta(a_i,-)$ where *?* is any symbol in the underlying alphabet **Ψ**. The beauty of the systolic array is that it can perform comparisons of **A** against a stream of **B** sequences, processing each symbol of the target sequences in constant time per symbol. With current technology, chips of this kind operate at rates of 3 million to 4 million symbols per second. A systolic array of 1,000 of these simple processors computes an aggregate of 3 billion to 4 billion dynamic programming entries per second.

COMPARING ONE SEQUENCE AGAINST A DATABASE

The current GENBANK database (Benson et al., 1993) of DNA sequences contains approximately 191 million nucleotides of sequence in about 183,000 sequence entries, and the PIR database (Barker et al., 1993) of protein sequences contains about 21 million amino acids of data in about 71,000 protein entries. Whenever a new DNA or protein sequence is produced in a laboratory, it is now routine practice to search these databases to see if the new sequence shares any similarities with existing entries. In the event that the new sequence is of unknown function, an interesting global or local similarity to an already-studied sequence may suggest possible functions. Thousands of such searches are performed every day.

In the case of protein databases, each entry is for a protein between 100 and 1,500 amino acids long, the average length being about 300. The entries in DNA databases have tended to be for segments of an organism's DNA that are of interest, such as stretches that code for proteins. These segments vary in length from 100 to 10,000 nucleotides. The limited length here is not intrinsic to the object as in the case of proteins, but because of limitations in the technology and the cost of obtaining long DNA sequences. In the early 1980s the longest consecutive stretches being sequenced were up to 5,000 nucleotides long. Today the sequences of some viruses of length 50,000 to 100,000 have been determined. Ultimately, what we will have is the entire sequence of DNA in a chromosome (100 million to 10 billion nucleotides), and entries in the database will simply be annotations describing interesting parts of these massive sequences.

A similarity search of a database takes a relatively short query sequence of a protein or DNA fragment and searches every entry in the database for evidence of similarity with the query. In protein databases, the query sequence and the entries in the database are typically of similar sizes. In DNA databases, the entries are typically much longer than the query sequence, and one is looking for subsegments of the entry that match the query.

Heuristic Algorithms

The problem of searching for *protein* similarities efficiently has led many investigators to abandon dynamic programming algorithms (for which the size of the problem has become too large) and instead consider designing very fast heuristic procedures: simple, often ad hoc, computational procedures that produce answers that are "nearly" correct with respect to a formally stated optimization criterion. One of the most popular database searching tools of this genre is FASTA (Lipman and Pearson, 1985). FASTA looks for entries that share a significant number of short identical subsequences of symbols with the query sequence. Any entry meeting this criterion is then compared via dynamic programming with the query sequence. In this way, the vast majority of entries are eliminated from consideration quickly. FASTA reports most of the alignments that would be identified by an equivalent dynamic programming calculation, but it misses some matches and also reports some spurious matches. On the other hand, FASTA is very fast.

A more recently developed heuristic algorithm is BLASTA (Altschul et al., 1990). BLASTA is faster than FASTA but is capable of detecting biologically meaningful similarities with accuracy comparable to that of FASTA. Given a query **A** and an entry **B**, BLASTA searches for *segment pairs* of high score. A segment pair is a substring from **A** and a substring from **B** of equal length, and the score of the pair is that of the no-gap alignment between them. One can argue that the presence of a high-scoring segment pair or pairs is evidence of functional similarity between proteins, because insertion and deletion events tend to significantly change the shape of a protein and hence its function. Note that segment pairs embody a local similarity concept. What is particularly useful is that there is a formula for the probability that two sequences have a segment pair above a certain score. Thus BLASTA can give an assessment of the statistical significance of any match that it reports. For a given threshold, T, BLASTA returns to the user all database entries that have a segment pair with the query of score greater than T ranked according to probability. BLASTA may miss some such matches, although in practice it misses very few.

The central idea used in BLASTA is the notion of a *neighborhood*. The t-neighborhood of a sequence **S** is the set of all sequences that align with **S** with score better than t. In the case of BLASTA, the t-neighborhood of **S** is exactly those sequences of equal length that form a segment pair of score higher than t under the Dayhoff scoring scheme (see Figure 3.5). This concept suggests a simple strategy for finding all entries that have segment pairs of length k and score greater than t with the query: generate the set of all sequences that are in the t-neighborhood of some k-substring of the query and see if an entry contains one of these strings. Scanning for an exact match to one of the strings in the neighborhood can be performed very efficiently: on the order of 0.5 million characters per second on a 20 SPECint computer. Unfortunately, for the general problem, the length of the segment pair is not known in advance, and even more devastating is the fact that the number of sequences in a neighborhood grows exponentially in both k and t, rendering it impractical for reasonable values of t. To circumvent this difficulty, BLASTA uses the fast scanning strategy above to find short segment pairs of length k above a score t, and then checks each of these to see if they are a portion of a segment pair of score T or greater. This approach is heuristic (that is, may miss some segment pairs) because it is possible for every length k subsegment pair of a segment pair of score T to have score less than t. Nonetheless, with $k = 4$ and $t = 17$ such misses are very rare, and BLASTA takes about 3 seconds for every 1 million characters of data searched.

To get an idea of the relative efficiency of various similarity searching approaches, consider the following rough timing estimates for a typical 20 SPECint workstation and a search against a typical protein query. The dynamic programming algorithm for local similarities presented above (also known as the Smith-Waterman algorithm) takes roughly 1000.0N microseconds to search a database with a total of N characters in it. On the other hand, FASTA takes 20.0N microseconds, and BLASTA only about 2.0N microseconds. At the other end of the spectrum, the systolic array chip described above takes only $0.3N$ microseconds to perform the Smith-Waterman algorithm with its special-purpose (and expensive) hardware.

Sublinear Similarity Searches

The total number N of characters of sequence in biosequence databases is growing exponentially. On the other hand, the size of the query sequences is basically fixed; for example, a protein sequence's length is bounded by 1,500 and averages 300. So designers of efficient computational methods should be principally concerned with how the time to perform such a search grows as a function of N. Yet all currently used methods take an amount of time that grows linearly in N; that is, they are $O(N)$ algorithms. This includes not only rigorous methods such as the dynamic programming algorithms mentioned above but also the popular heuristics FASTA and BLASTA. Even the systolic array chips described above do not change this. When a database increases in size by a factor of 1,000, all of these $O(N)$ methods take 1,000 times longer to search that database. Using the timing estimates given above, it follows that while a custom chip may take about 3 seconds to search 10 million amino acids or nucleotides, it will take 3,000 seconds, or about 50 minutes, to search 10 billion symbols. And this is the fastest of the linear methods: BLASTA will take hours, and the Smith-Waterman algorithm will take months. One could resort to massive parallelism, but such machinery is beyond the budget of most investigators, and it is unlikely that speedups due to improvements in hardware technology will keep up with sequencing rates in the next decade.

What would be very desirable, if not essential, is to have search methods with computing time sublinear in N, that is, $O(N^{\alpha})$ for some $\alpha < 1$. For example, suppose there is an algorithm that takes $O(N^{0.5})$ time, which is to say that as N grows, the time taken grows as the square root of N. For example, if the algorithm takes about 10 seconds on a 10 million symbol database, then on 10 billion symbols, it will take about $1{,}000^{0.5} \approx 31$ times longer, or about 5 minutes. Note that while an $O(N^{0.5})$ algorithm may be slower than an $O(N)$ algorithm on 10 million symbols, it may be faster on 10 billion. Figure 3.9 illustrates this "crossover": in this figure, the size of N at which the $O(N^{0.5})$ algorithm overtakes the $O(N)$ algorithm is approximately 1×10^{8}. Similarly, an $O(N^{0.2})$ algorithm that takes, say, 15 seconds on 10 million symbols, will take about 1 minute, or only 4 times longer, on 10 billion. To forcefully illustrate the point, we chose to let our examples be slower at $N = 10$ million than the competing $O(N)$ algorithm. As will be seen in a moment, a sublinear algorithm does exist that is actually already much faster on databases of size 10 million. The other important thing to note is that we are not considering heuristic algorithms here. What we desire is nothing less than algorithms that accomplish exactly the same computational task of complete comparison as the dynamic programming algorithms, but are much faster because the computation is performed in a clever way.

A recent result on the approximate string matching problem under the simple unit-cost scheme of Figure 3.1 portends the possibility of truly sublinear algorithms for the general problem. For relatively stringent matches, this new algorithm is 3 to 4 orders of magnitude more efficient

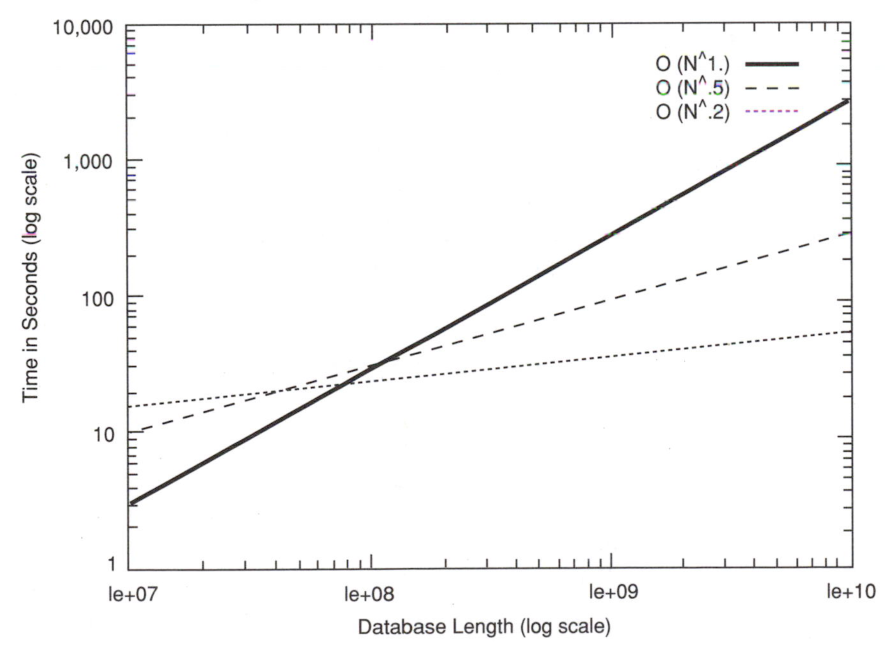

FIGURE 3.9 Sublinear versus linear algorithms.

than the equivalent dynamic programming computation on a database of 1 million characters. On the other hand, the approximate string matching problem is a special case of the more biologically relevant computation that involves more general scoring schemes such as the ones in Figures 3.4 and 3.5, and a sublinear algorithm for the general problem has yet to be achieved.

We conclude with a few more details on this sublinear algorithm (Myers, 1994b). For a search of matches to a query of length P with D or fewer differences, the quantity $\varepsilon = D/P$ is the maximum fraction of differences permitted per unit length of the query and is called the *mismatch ratio*. Searching for such an approximate string match over a database of length N can be accomplished in $O(DN^{pow(\varepsilon)} \log N)$ expected time with the new algorithm. The exponent is an increasing and concave function of ε that is 0 when $\varepsilon = 0$ and depends on the size $|\Psi|$ of the underlying alphabet. The algorithm is superior to the $O(N)$ algorithms and truly sublinear in N when ε is small enough to guarantee that $pow(\varepsilon) < 1$. For example, $pow(\varepsilon)$ is less than 1 when $\varepsilon < 33$ percent for $|\Psi| = 4$ (DNA alphabet) and when $\varepsilon <$ 56 percent for $|\Psi| = 20$ (protein alphabet). More specifically, $pow(\varepsilon) \le 0.22 + 2.3\varepsilon$ when $|\Psi| = 4$ and $pow(\varepsilon) \le 0.17 + 1.4\varepsilon$ when $|\Psi| = 20$. So, for DNA, the algorithm takes a maximum of $O(N^{0.5})$ time when ε is 12 percent, and for proteins, a maximum of $O(N^{0.5})$ time when ε is 26 percent. The logic used to prove these bounds is coarse, and, in practice, the performance of these methods is much better than the bounds indicate. If these results can be extended to handle the more general problem of arbitrary scoring tables, the impact on the field could be great.

OPEN PROBLEMS

While progress is continually being made on existing problems in sequence comparison, new problems continue to arise. A fundamental issue is the definition of similarity. We have focused here only on the insertion-deletion-substitution model of comparison and some small variations. Some authors (e.g., Altshul and Erikson, 1986) have looked at nonadditive scoring schemes that are intended to reflect the probability of finding a given alignment by chance. A fundamental change in the optimization criterion for alignment creates a new set of algorithmic problems.

What about fundamentally speeding up sequence comparisons? The best lower bounds placed on algorithms for comparing two sequences of length N is $O(N \log N)$, yet the fastest algorithm takes $O(N^2 / \log^2 N)$ time (Masek and Paterson, 1980). Can this gap be narrowed, either from above (finding faster algorithms) or below (finding lower bounds that are higher)? Can we perform faster database searches for the case of generalized Levenshtein scores, as is suggested by the results given above for the approximate string matching problem? Speeding up database searches is very important. Are there other effective ways to parallelize such searches or to exploit preprocessing of the databases, such as an index?

Biologists are interested in searching databases for patterns other than given strings or regular expressions. Recently, fast algorithms have been developed (Kannan and Myers, 1993; Landau and Schmidt, 1993) for finding approximate repeats, for example, finding a pattern that matches some string **X** and then 5 to 10 symbols to the right matches the same string modulo 5 percent differences. Many DNA structures are induced by forming base pairs that can be viewed as finding approximate palindromes separated by a given range of spacing. More intricate patterns for protein motifs and secondary structure are suggested by the systems QUEST (Arbarbanel et al., 1984), ARIADNE (Lathrop et al., 1987), and ANREP (Mehldau and Myers, 1993), all of which pose problems that could use algorithmic refinement.

Finally, biologists compare objects other than sequences. For example, the partial sequence information of a restriction map can be viewed as a string on which one has placed a large number of beads of, say, eight colors, at various positions along the string. Given two such maps, are they similar? This problem has been examined by several authors (e.g., Miller et al., 1990). There are still fundamental questions as to what the measure of similarity should be and how to design efficient algorithms for each. There has also been work on comparing phylogenetic trees and chromosome staining patterns (e.g., Zhang and Shasha, 1989). Indubitably the list will continue to grow.

REFERENCES

Altschul, S.F., and B.W. Erikson, 1986, "Locally optimal subalignments using nonlinear similarity functions," *Bull. Math. Biol.* **48**(5/6), 633-660.

Altschul, S.F., W. Gish, W. Miller, E.W. Myers, and D.J. Lipman, 1990, "A basic local alignment search tool," *Journal of Molecular Biology* **215**, 403-410.

Arbarbanel, R.M., P.R. Wieneke, E. Mansfield, D.A. Jaffe, and D.L. Brutlag, 1984, "Rapid searches for complex patterns in biological molecules," *Nucleic Acids Research* **12**(1), 263-280.

Barker, W.C., D. George, H.-W. Mewes, F. Pfeiffer, and A. Tsugita, 1993, "The PIR-International database," *Nucleic Acids Research* **21**(13), 3089-3092.

Benson, D., D.J. Lipman, and J. Ostell, 1993, "GenBank," *Nucleic Acids Research* **21**(13), 2963-2965.

Chao, K.-M., and W. Miller, 1994, "Linear-space algorithms that build local alignments from fragments," *Algorithmica*, in press.

Dayhoff, M.O., W.C. Barker, and L.T. Hunt, 1983, "Establishing homologies in protein sequences," *Methods in Enzymology* **91**, 524-545.

Eppstein, D., Z. Galil, and R. Giancarlo, 1989, "Speeding up dynamic programming," *Theoretical Computer Science* **64**, 107-118.

Feng, D.F., and R.F. Doolittle, 1987, "Progressive sequence alignment as a prerequisite to correct phylogenetic trees," *Journal of Molecular Evolution* **25**, 351-360.

Fox, B., 1973, "Calculating the Kth shortest paths," *INFOR J (Can. J. Oper. Res. Inf. Process.)* **11**, 66-70.

Garey, M.R., and D.S. Johnson, 1979, *Computers and Intractability: A Guide to the Theory of NP-Complete Problems*, New York: W.H. Freeman Press.

Gotoh, O., 1982, "An improved algorithm for matching biological sequences," *Journal of Molecular Biology* **162**, 705-708.

Horowitz, E., and S. Sahni, 1978, pp. 198-247 in *Fundamentals of Computer Algorithms*, New York: Computer Science Press.

Kannan, S.K., and E.W. Myers, 1993, "An algorithm for locating non-overlapping regions of maximum alignment score," *Proceedings of the 4th Combinatorial Pattern Matching Conference*, Springer-Verlag Lecture Notes in Computer Science 684, 74-86.

Landau, G.M., and J.P. Schmidt, 1993, "An algorithm for approximate tandem repeats," *Proceedings of the 4th Combinatorial Pattern Matching Conference*, Springer-Verlag Lecture Notes in Computer Science 684, 120-133.

Lathrop, R.H., T.A. Webster, and T.F. Smith, 1987, "ARIADNE: A flexible framework for protein structure recognition," *Commun. ACM* **30**, 909-921.

Lipman, D.J., and W.R. Pearson, 1985, "Rapid and sensitive protein similarity searches," *Science* **227**, 1435-1441.

Masek, W.J., and M.S. Paterson, 1980, "A faster algorithm for computing string-edit distances," *Journal of Computing Systems Science* **20**(1), 18-31.

Mehldau, G., and E.W. Myers, 1993, "A system for pattern matching applications on biosequences," *CABIOS* **9**, 3, 299-314.

Miller, W., and E.W. Myers, 1988, "Sequence comparison with concave weighting functions," *Bull. Math. Biology* **50**(2), 97-120.

Miller, W., J. Ostell, and K.E. Rudd, 1990, "An algorithm for searching restriction maps," *CABIOS* **6,** 247-252.

Myers, E.W., 1994a, "Approximately Matching Context Free Languages," TR94-22, Department of Computer Science, University of Arizona, Tucson, Ariz.

Myers, E.W., 1994b, "A sublinear algorithm for approximate keywork searching," *Algorithmica* **12**(4), 345-374.

Myers, E.W., and W. Miller, 1988, "Optimal alignments in linear space," *CABIOS* **4**(1), 11-17.

Myers, E.W., and W. Miller, 1989, "Approximate matching of regular expressions," *Bull. Math. Biol.* **51**(1), 5-37.

Sankoff, D., 1975, "Minimal mutation trees of sequences," *SIAM Journal of Applied Mathematics* **28**(1), 35-42.

Smith, T.F., and M.S. Waterman, 1981, "Identification of common molecular sequences," *Journal of Molecular Biology* **147**, 195-197.

Smith, T.F., and W.S. Fitch, 1983, "Optimal sequence alignments," *Proceedings of the National Academy of Sciences USA* **80,** 1382-1386.

Waterman, M.S., and M. Eggert, 1987, "A new algorithm for best subsequence alignments with application to tRNA-rRNA comparisons," *Journal of Molecular Biology* **197**, 723-728.

Waterman, M.S., M. Eggert, and E. Lander, 1992, "Parametric sequence comparisons," *Proceedings of the National Academy of Sciences USA* **89**, 6090-6093.

Waterman, M.S., T.F. Smith, and W.A. Beyer, 1976, "Some biological sequence metrics," *Advances in Mathematics* **20**, 367-387.

Zhang, K., and D. Shasha, 1989, "Simple fast algorithms for the editing distance between trees and related problems," *SIAM Journal on Computing* **18**, 1245-1262.

Chapter 4
Hearing Distant Echoes: Using Extremal Statistics to Probe Evolutionary Origins

Michael S. Waterman
University of Southern California

The comparison of DNA and protein sequences provides a powerful tool for discerning the function, structure, and evolutionary origin of important macromolecules. Sequence comparison sometimes reveals striking matches between molecules that were hitherto not known to be related—immediately suggesting hypotheses that can be tested in the laboratory. In other cases, sequence comparison reveals only weak similarities. In such instances, statistical theory is essential for interpreting the significance of such matches. The author discusses large deviation theory for sequence matching and applies it to evaluate a tantalizing report concerning distant echoes from the earliest period in the origin of life.

As soon as new deoxyribonucleic acid (DNA) or protein sequences are determined, molecular biologists immediately examine them for clues about their biological significance. A number of important questions about the function of a newly determined protein are often asked, including the following. What can be inferred about the function of a new protein on the basis of its amino acid sequence? Can one discern the reactions it catalyzes or the molecules it binds? What three-dimensional shape will the linear amino acid sequence of a protein assume when it folds up according to the laws of thermodynamics? Another class of questions concerns the evolutionary relationships between known sequences. For example, some questions concerning hemoglobin are the following. What is the evolutionary relationship between three related α, β, and γ hemoglobin genes? What is the evolutionary relationship between the hemoglobin molecules from various organisms? What do these sequences tell us about the evolutionary history of humans, chimpanzees, and gorillas? Each of these questions can be approached, if not always entirely solved, by sequence comparison.

Sequence comparison is of tremendous interest to molecular biologists because it is becoming easy to determine DNA and protein sequences, whereas it remains difficult to determine molecular structure or function by experimental means. Thus, functional and structural clues from sequence analysis can save years of work at the laboratory bench. An important early example illustrates the point. Some years ago, molecular biologists compared the protein sequence encoded by a cancer-causing gene (or oncogene) called *v-sis* to the available database of protein sequences. Remarkably, a computer search revealed that the sequence showed more than 90 percent identity to the sequence of a previously discovered gene encoding a growth-stimulating molecule, called platelet-derived growth factor (PDGF). Instantly, cancer researchers had a precise hypothesis about how this oncogene causes unregulated cell growth. Subsequent experiments confirmed the guess.

Nowadays, molecular biologists routinely carry out such computer searches against the current databases (which now contain both protein and DNA sequences) and are rewarded with striking and suggestive matches at a high frequency (perhaps 20 to 30 percent for a new gene). In some cases, the matches extend across the entire length of the protein. In other cases, there is a strong match across a restricted domain—examples include particular sequences at the catalytic site of enzymes that hydrolyze adenosinetriphosphate (ATP) or at the DNA-binding site of proteins that regulate the activity of genes. The frequency with which such strong matches are found is a tribute to the tremendously conservative nature of evolution: many of the basic building blocks of proteins and DNA have been reused in hundreds of different ways.

For the majority of new sequences, however, there is no striking match in the database. Although this may change with time (some molecular biologists believe that there are only a few thousand or a few tens of thousands of basic architectural motifs for proteins and that it is just a matter of time before we collect them all), computer searches will turn up only weak similarities. Before attempting to read biological significance into such weak similarities, one must evaluate their statistical significance. Not surprisingly, this is an area in which mathematics has much to offer molecular biology. To motivate the study of the statistical significance of sequence similarities, we consider a single data set that provoked a great deal of excitement a few years ago when a team of researchers thought that they saw extraordinary clues about early evolution in the sequences of genes encoding certain ribonucleic acid (RNA) molecules.

The origin of the universe and the origin of life are topics of wide interest to both biologists and nonbiologists. One approach to studying the origin of the universe is to listen to faint echoes from the Big Bang. Similar approaches are used in studying the origin of life. Are there any molecular echoes remaining from the origin of life? Each of the three key molecules in molecular biology—DNA, RNA, and protein—has been championed by some theorists as the earliest self-replicating molecule. Proteins have seemed attractive to some because of their ability to catalyze chemical reactions. DNA has seemed attractive to others because it is a stable store of information. Lately, however, RNA has taken the lead based on the well-known ability of RNA to encode information in the same manner as DNA and the recently discovered ability of RNA to act as nonprotein enzymes that are able to catalyze some chemical reactions. These properties suggest that some RNA sequence might have been able to achieve the key feat of self-replication—serving as both self-template and replication enzyme. Thus, life may have started out as an RNA world.

As indicated in Chapter 1, modern RNAs come in three varieties: messenger RNAs (mRNAs), ribosomal RNAs (rRNAs), and transfer RNAs (tRNAs). mRNAs are the messages copied from genes. rRNAs are components of the macromolecular structure, called the ribosome, used for translating RNA sequences into protein sequences. tRNAs are the "adapter molecules" that read the genetic code, with an anticodon loop recognizing a particular codon at one end and an attachment site for the amino acid corresponding to this codon at the other. rRNAs and tRNAs are clearly ancient inventions, necessary for the progression from life based only on RNA to organisms employing proteins for efficient catalysis of biochemical reactions.

In the early 1980s David Bloch and colleagues reported that they had found that these two types of RNA—tRNA and rRNA—had significant sequence similarities implying a common evolutionary ancestry (Bloch et al., 1983). In his paper, Bloch reported:

> Many tRNAs of *E. coli* and yeast contain stretches whose base sequences are similar to those found in their respective rRNAs. The matches are too frequent and extensive to be attributed to coincidence. They are distributed without discernible pattern along and among the RNAs and between the two species. They occur in loops as well as in stems, among both conserved and non-conserved regions. Their distributions suggest that they reflect common ancestral origins rather than common functions, and that they represent true homologies.

Such tantalizing arguments should be grounded in statistics—since we cannot test the origin of life by direct experiment (as we could test a proposed function for a protein based on sequence similarity). In this chapter, we develop some tools for evaluating statistical significance and apply them to Bloch's data.

The biological hypothesis that relationships between the RNAs are true homologies is necessarily imprecise. The evidence given is frequent and extensive matchings of stretches of sequences between the molecules, just the sort of matchings that the local algorithm presented below is designed to find. To "test" the biological hypothesis, we form a statistical hypothesis that the sequences are generated with independent and identically distributed letters. Then we test this hypothesis by computing scores using the local algorithm. Since letters in real sequences are not independent, it is possible to change the hypothesis to a Markov hypothesis, for example. This does not change the score distribution very much for the distributions obtained from real sequences. If the score distribution is consistent with that from comparison of random sequences, we would fail to reject the statistical hypothesis and thus have evidence against the biological hypothesis. If, on the other hand, the scores are frequently too large, showing strongly matching stretches or intervals of sequence, we have evidence for the biological hypothesis and against the statistical hypothesis.

Statistical questions are increasingly important in molecular biology. While statistical significance is not directly related to biological importance, it is a good indicator and can lead to the formulation of important biological hypotheses, as noted above. Conversely, lack of statistical significance is an important clue in considering whether to reject a relationship that may seem interesting to the human eye. With over 70,000 sequences in modern databases, molecular biologists require an automatic way to reject all but the most interesting results from a database search. Comparing one sequence to the database involves 70,000 comparisons. Comparing all pairs of sequences involves $\binom{70{,}000}{2}$, or about 2.4×10^9, comparisons. As we will see with the tRNA and rRNA comparison, even a small number of comparisons can raise subtle questions.

GLOBAL SEQUENCE COMPARISONS

We will now discuss a number of situations for sequence comparisons and some probability and statistics that can be applied to these problems. Some powerful and elegant mathematics has been developed to treat this class of problems. Our discussion will naturally break into two parts, global comparisons and local comparisons.

Sequence Alignment

In this section we study the comparison of two sequences. For simplicity the two sequences $A_1 A_2 \ldots A_n$ and $B_1 B_2 \ldots B_m$ will consist of letters drawn independently with identical distribution from a common alphabet.

Sequences evolve at the molecular level by several mechanisms. One letter, `A` for example, can be substituted for another, `G` for example. These events are called substitutions. Letters can be removed from or added to a sequence, and these events are called deletions or insertions. Given two sequences such as `ATTGCC` and `ACGGC`, it is usually not clear how they should be related. The possible relationships are often written as alignments such as:

```
ATTGCC
ACGG-C
```

or

```
ATTGCC
-ACGGC
```

where in the first case there are three identities, two substitutions, and one insertion/deletion (indel) and in the second case there are two identities, three substitutions, and one indel.

An alignment can be obtained by inserting gaps ("–") into the sequences so that

$$A_1 A_2 \ldots A_n \rightarrow A_1^* A_2^* \ldots A_L^*$$

and

$$B_1 B_2 \ldots B_m \rightarrow B_1^* B_2^* \ldots B_L^* .$$

Here the subsequence of all $A_i^* \neq$ "–" is identical to $A_1 A_2 \ldots A_n$. Then, since the *-sequences have equal length, A_i^* is aligned with B_i^*. In Chapter 3, algorithms to achieve optimal alignments are discussed. Here we are interested in the statistical distribution of these scores, not in how they are obtained. Global alignments refer to the situation where all the letters of each sequence must be accounted for in the alignments. There are two types of global alignments, alignments where the pairing is given and alignments where the pairing is not given.

Alignment Given

In this section, we assume the alignment is given with the sequences:

$$\begin{matrix} A_1 A_2 \ldots A_n \\ B_1 B_2 \ldots B_n . \end{matrix}$$

(Gaps "–" have been added so that these sequences both have the same length—L in the previous section, n here—and the stars have been omitted to simplify the notation.) In this case the alignment is given and therefore cannot be optimized. We give the statistical distribution of the alignment score for completeness, however. Let $s(A,B)$ be a real valued random variable. Define the score S by

$$S = \sum_{i=1}^{n} s(A_i, B_i),$$

and let $\mathbf{E}(S)$ denote the expectation of S and Var (S) denote the variance. Clearly, $\mathbf{E}(S) = n\mathbf{E}(s(A,B))$ and

$$\mathrm{Var}(S) = n\,\mathrm{Var}(s(A,B)).$$

Since S is the sum of independent, identically distributed random variables $s(A,B)$, the central limit theorem implies that for large n

$$S \approx \mathrm{Normal}(n\,\mathbf{E}(s(A,B)), n\mathrm{Var}(s(A,B))).$$

If $s(A,B) \in \{0,1\}$, then S is binomial (n,p), where $p = \mathbf{P}\{s(A,B)=1\}$. Even when the letters are not identically distributed, the limiting distribution is normal under mild assumptions (Chung, 1974).

Alignment Unknown

The assumptions of the last section are carried over: $A_1 A_2 \ldots A_n$ and $B_1 B_2 \ldots B_m$ are composed of independent and identically distributed letters and $s(A,B)$ is a real valued random variable on pairs of letters. We extend $s(\cdot,\cdot)$ to $s(A,-)$ and $s(-,B)$ so that deletions are included. We assume that the value of s for all deletion scores is smaller than $\max s(A,B)$. An alignment score S is the maximum over all possible alignments

$$S = \max_{\text{alignments}} \sum_{i=1}^{L} s(A_i^*, B_i^*).$$

The optimization destroys the classical normal distribution of alignment score, but an easy application of a beautiful theorem known as Kingman's subadditive ergodic theorem gives an interesting result:

Theorem 4.1 (Kingman, 1973) *For s,t nonnegative integers with $0 \le s \le t$, let $X_{s,t}$ be a collection of random variables that satisfy*

(i) Whenever $s < t < u$, $X_{s,u} \le X_{s,t} + X_{t,u}$,

(ii) The joint distribution of $\{X_{s,t}\}$ is the same as that of $\{X_{s+1,t+1}\}$,

(iii) The expectation $g_t = \mathbf{E}(X_{0,t})$ exists and satisfies $g_t \ge -Kt$ for some constant K and all $t > 1$.

Then the finite $\lim_{t\to\infty} X_{0,t} / t = \lambda$ exists with probability 1 *and in the mean.*

The essential assumption is (ii), the subadditivity condition. To motivate and illustrate this theorem, recall the strong law of large numbers (SLLN), which treats independent, identically distributed (iid) random variables $W_1, W_2, \ldots$ with $\mu = \mathbf{E}(W_i)$. The SLLN asserts that

$$\frac{W_1 + W_2 + \ldots + W_n}{n} \to \mu$$

with probability 1.

It is easy to see that additivity holds. Set

$$U_{s,t} = \sum_{s+1 \le i \le t} W_i .$$

Of course (i) is satisfied:

$$U_{s,u} = \sum_{s+1 \le i \le t} W_i + \sum_{t+1 \le i \le u} W_i$$

$$= U_{s,t} + U_{t,u} .$$

Since the W_i are iid, (ii) is evidently true. Finally, $g(t) = \mathbf{E}(U_{0,t}) = t\mu$, so that (iii) holds with $\mu = -K$. Therefore the limit

$$\lim_{t \to \infty} \sum_{1 \le i \le t} W_i / t$$

exists and is constant with probability 1. Notice that this setup does not allow us to conclude that the limit is μ. This is a price of relaxing the assumption of additivity.

Returning to the statistical distribution of alignment score, recall that an alignment score S is the maximum over all possible alignments

$$S = \max_{\text{alignments}} \sum_{i=1}^{L} s(A_i^*, B_i^*) .$$

Define $X_{s,t}$ by

$$-X_{s,t} = \text{score of } A_{s+1} A_{s+2} \ldots A_t \text{ vs } B_{s+1} B_{s+2} \ldots B_t .$$

Then evidently,

$$-X_{s,u} \ge (-X_{s,t}) + (-X_{t,u})$$

and

$$X_{s,u} \le X_{s,t} + X_{t,u} .$$

We have that $g_t = \mathbf{E}\left(X_{0,t}\right)$ exists since the expectation of a single alignment exists and $-X_{0,t}$ is the maximum of a *finite* number of alignment scores. The final hypothesis to check is $g_t \ge -Kt$ for some constant K and all $t > 1$. Clearly,

$$\mathbf{E}\left(-X_{0,t}\right) \le t \max s(A, B)$$

so that

$$g_t \ge -(\max s(A, B))t = -Kt .$$

Our conclusion is that

$$\lim_{t \to \infty} X_{0,t} / t = \lambda$$

exists with probability 1 and in the mean. Therefore optimal alignment score grows linearly with sequence length. Obviously, $\lambda \geq \mathbf{E}(s(A, B))$.

In the simplest case of interest, the alphabet has two uniformly distributed letters and $s(A, B) = 0$ if $A \neq B$ and $s(A, A) = s(B, B) = 1$. The alignment score is known as the *longest common subsequence*, and Chvátal and Sankoff (1975) wrote a seminal paper on this problem in the 1970s. In spite of much effort since then, λ remains undetermined. Deken (1979) gives bounds for λ: $0.7615 \leq \lambda \leq 0.8602$. Without alignment the fraction of matching letters is 0.5 by the strong law of large numbers. Not too much is known about the variance either, although Steele (1986) proves it is $O(n)$.

LOCAL SEQUENCE COMPARISONS

Alignment Given

Consider many independent throws of a coin with probability p of heads, where $0 < p < 1$. For any p, there will be stretches where the coin comes up heads every time. What is the distribution of the length of the longest of these head runs? This maximum length is known as a "local" score; while it is a global maximum, it is a function only of the nearby tosses. A related problem is to consider sequence $A_1 A_2 \ldots A_n$ of letters chosen independently and from a common alphabet, {`A`, `C`, `G`, `T`} for DNA for example. The letters `A` and `G` are known as purines (R), and `C` and `T` are known as pyrimidines (Y). A two-letter alphabet is natural when grouping nucleotides by chemical similarity. In fact, there is a hypothesis that the first nucleic acid sequences were made up of just two elements, R and Y. It is natural to ask how random the distribution of R and Y is for a given sequence. We will study how large is R_n, the length of the longest run of purines R in a sequence of length n. Here an occurrence of R is a "head" for the coin tossing analogy.

The coin tossing question was considered by Erdös and Rényi (1970), who found the strong law

$$\lim_{n\to\infty} \frac{R_n}{\log_{1/p} n} = 1 \text{ with probability 1.} \tag{4.1}$$

Of course, one may desire more detailed information about R_n. For an example, we look at "16**S**" rRNA from *E. coli,* which is 1,541 letters in length and is known by its sedimentation rate **S** of 16. (The sedimentation rate is an indication of mass: the greater the mass, the higher the rate of sedimentation and the larger the value of **S**.) Equation (4.1) tells us that we would typically see about $\log_{1/p} 1{,}541 = 10.6$ R's in a row where $p = \frac{1}{2}$. What if we have a head run of length 14 in our 16**S** sequence? Is this score extreme enough to be of note? For the statistical question of significance, we need to have a way to compute such probabilities. This is supplied by Poisson approximation.

For an appropriately chosen test length t, we see an R run of length t begins at a given position with a small probability. Since the number of positions where such a run could occur is large, the number of long head runs should be approximately Poisson. Our discussion about the mathematics behind this intuition follows Goldstein (1990).

This intuition is almost correct. One must first, however, adjust for the fact that runs of heads occur in "clumps"; that is, if there is a run of heads of length t beginning at position α, then with probability p there will also be a run of heads of length t beginning at position $\alpha + 1$, with probability p^2 a run of heads of length t beginning at position $\alpha + 2$, and so forth. Hence, the total number of runs of length t or more is seen to have a compound Poisson distribution. By counting only the first such run in every clump, the occurrences now counted are no longer clumped and their number is approximately Poisson. This is an example, with average clump size $1 + p + p^2 \ldots = 1/(1-p)$, of the "Poisson clumping heuristic," as described by Aldous (1989). Using the fact that having no runs of length t is equivalent to having the longest head run shorter than t, we can approximate the distribution of the length of the longest run of heads. In the remainder of this section we explore the approximation of this distribution by the Poisson distribution.

Let I be an index set, and for each $\alpha \in I$, let X_α be an indicator random variable, that is, $X_\alpha = 1$ if an event occurs and $X_\alpha = 0$ if the event does not occur. The total number of occurrences of events can be expressed as

$$W = \sum_{\alpha \in I} X_\alpha .$$

It seems intuitive that if $p_\alpha = \mathbf{P}(X_\alpha = 1)$ is small, and $|I|$, the size of the index set, is large, then W should be approximately Poisson distributed. Certainly this is true when all the $X_\alpha, \alpha \in I$, are independent. In the case of dependence, it seems plausible that the same approximation should hold when dependence is somewhat confined. For each α, we let B_α be the set of dependence for α; that is, for each $\alpha \in I$, assume we are given a set $B_\alpha \subset I$ such that

$$X_\alpha \text{ is independent of } X_\beta, \beta \notin B_\alpha . \tag{4.2}$$

Define

$$b_1 \equiv \sum_{\alpha \in I} \sum_{\beta \in B_\alpha} p_\alpha p_\beta \quad \text{and}$$
$$b_2 \equiv \sum_{\alpha \in I} \sum_{\alpha \neq \beta \in B_\alpha} p_{\alpha\beta}, \quad \text{where } p_{\alpha\beta} \equiv \mathbf{E}\left(X_\alpha X_\beta\right).$$

Let Z denote a Poisson random variable with mean λ, so that for $k = 0, 1, 2, \ldots$,

$$\mathbf{P}(Z = k) = e^{-\lambda} \frac{\lambda^k}{k!} .$$

Classically, the Poisson distribution is the probability law of rare events. It is remarkable that a few probability distributions arise with great frequency. The three principal distributions are the binomial, the normal, and the Poisson. (See Feller (1968) for an extensive discussion of these matters.)

Let $h: Z^+ \to R$, where $Z^+ = \{0,1,2,\ldots\}$, and $\|h\| \equiv \sup_{k\geq 0}|h(k)|$. We denote the total variation distance between the distributions of W and Z by

$$\begin{aligned}\|W - Z\| &\equiv \sup_{\|h\|=1} \left| \mathbf{E}(h(W)) - \mathbf{E}(h(Z)) \right| \\ &= 2 \sup_{A\subset Z^+} \left| \mathbf{P}(W \in A) - \mathbf{P}(Z \in A) \right|.\end{aligned}$$

More general versions of the following theorem appear in Arratia et al. (1989, 1990). They refer to this approach as the Chen-Stein method.

Theorem 4.2 *Let W be the number of occurrences of dependent events, and let Z be a Poisson random variable with* $\mathbf{E}(Z) = \mathbf{E}(W) = \lambda$. *Then*

$$\|W - Z\| \leq 2(b_1 + b_2)\frac{1 - e^{-\lambda}}{\lambda} \leq 2(b_1 + b_2),$$

and in particular

$$\left|\mathbf{P}(W = 0) - e^{-\lambda}\right| \leq (b_1 + b_2)(1 - e^{-\lambda}) / \lambda .$$

We first apply Poisson approximation to the distribution of the length of long success runs in Bernoulli trials. This has application to molecular sequences and provides a good illustration of the methods needed for sequence comparisons in the case when the alignment is unknown. Let $C_1, C_2, \ldots$ be independent Bernoulli random variables with success probability p, and let R_n be the length of the longest run of heads contained in the first n tosses. Fix a test level t and let the index set be $I = \{1, 2, \ldots, n - t + 1\}$; the elements of the index set will denote locations where head runs of length t or greater may begin. A head run of length t or more begins at position 1 if and only if the indicator random variable

$$X_1 = C_1 C_2 \ldots C_t$$

takes on the value 1. Now, to unclump the remaining runs define

$$X_\alpha = (1 - C_{\alpha-1})C_\alpha C_{\alpha+1} \ldots C_{\alpha+t-1} \text{ for } \alpha = 2,3,\ldots,n-t+1.$$

For $\alpha = 2,3,\ldots,n-t+1$, X_α will be 1 if and only if a run of t or more heads begins at position α, preceded by a tail. Below we calculate b_1, show that $b_2 = 0$, and find a bound for the approximation.

Write now the total number of clumps of runs of length t or more as the sum of dependent indicator random variables

$$W = \sum_{\alpha \in I} X_\alpha .$$

The Poisson approximation heuristic says we should be able to approximate the distribution of W by a Poisson random variable with mean

$$\lambda_n(t) = \mathbf{E}(W) = p^t\left((n-t)(1-p)+1\right). \tag{4.3}$$

In particular then, since we have as events

$$\{R_n < t\} = \{W = 0\},$$

the distribution function of R_n can be approximated as

$$\mathbf{P}(R_n < t) = \mathbf{P}(W = 0) \cong e^{-\lambda_n(t)}.$$

The test length t is dictated by requiring λ to be bounded away from 0 and ∞; this is equivalent to the condition that $t - \log_{1/p} n$ is bounded. In fact, for integer t, with c defined by

$$t = \log_{1/p}((n-t)(1-p)+1) + c,$$

the above approximation predicts that

$$\mathbf{P}(R_n < t) \cong e^{-\lambda_n(t)} = \exp(-p^c),$$

that is, that $R_n - \log_{1/p}((n-t)(1-p)+1)$ has an asymptotic extreme value distribution. This is almost so; the limiting distribution is complicated by the fact that R_n can assume only integer values. However, this fact does not complicate the approximation itself.

For an example we return to our problem with the 16**S** rRNA sequence. We model an R run of length 14 by $n = 1{,}541$ independent tosses of a fair coin. Using formula (4.3), we calculate that $\lambda_n = 0.0700$. Using the Poisson distribution, $\mathbf{P}(R_{1541} \geq 14)$ is approximately $1 - \exp(-\lambda_n) = 0.0676$.

Even so, without a bound on the error we have no way of knowing if the event is likely or not. To assess the accuracy of the above approximation, we apply Theorem 4.2. We define $B_\alpha = \{\beta \in I: |\alpha - \beta| \leq t\}$ for all α. Since X_α is independent of $\{X_\beta : \beta \notin B_\alpha\}$, condition 1 is satisfied. Furthermore, if $1 \leq |\alpha - \beta| \leq t$, we cannot have both X_α and X_β equal to 1 since we require that a run begin with a tail. Therefore $p_{\alpha\beta} = 0$ for $\beta \in B_\alpha, \beta \neq \alpha$, and hence $b_2 = 0$.

In order to calculate $b_1 = \sum_\alpha \sum_{\beta \in B_\alpha} p_\alpha p_\beta$, we break up the sum over $\beta \in B_\alpha$ into two parts, depending on whether or not p_1 appears. This yields the bound

$$b_1 < \lambda^2(2t+1)/(n-t+1) + 2\lambda p^t. \tag{4.4}$$

Theorem 4.2 now shows us that the Poisson approximation is quite accurate for the example considered above; the probability computed is correct to within $b_1 < 10^{-4}$, that is,

$$0.0699 \leq \mathbf{P}(R_n \geq 13) \leq 0.0701.$$

Alignment Unknown

The situation for matching between two sequences is closely related, although the dependence structure becomes more complex. Suppose that the two sequences $A_1A_2\ldots A_n$ and $B_1B_2\ldots B_m$ are made up of letters independently and uniformly chosen from a d-letter alphabet. It must be emphasized that whenever the letters are not uniformly chosen, Theorem 4.2 holds but is not straightforward to apply. In matching DNA, $d = 4$; for protein sequences, $d = 20$. Let

$$I = \{(i,j): 1 \le i \le n-t+1, 1 \le j \le m-t+1\}.$$

Define indicator random variables

$$E_{i,j} = 1 \quad \text{if} \quad A_i = B_j.$$

Let $p = \mathbf{P}(E_\alpha = 1) = 1/d$.

As in the case of head runs, we need to unclump matches and consider "boundary effects." Let

$$X_{i,j} = E_{i,j}E_{i+1,j+1}\cdots E_{i+t-1,j+t-1} \quad \text{if } i = 1 \text{ or } j = 1$$

and otherwise

$$X_{i,j} = (1 - E_{i-1,j-1})E_{i,j}E_{i+1,j+1}\cdots E_{i+t-1,j+t-1}.$$

With $W = \sum_{\alpha \in I} X_\alpha$, calculating $\lambda = \mathbf{E}(W)$ yields

$$\lambda = p^t[(n+m-2t+1)+(n-t)(m-t)(1-p)]. \qquad (4.5)$$

In matching two tRNA sequences, one of length 76, the other of length 77, would a match of length 9 be unusual? For the given parameters, $\lambda = 0.0136$ and under the model above, the event has a probability of approximately

$$1 - \exp(-\lambda) = 0.135.$$

A bound on the error may be calculated in a way similar to that for coin tossing. We note that again $b_2 = 0$, and by breaking the sum for b_1 into two sums, one of which is made up of all terms that involve the boundary, we find

$$b_1 < \lambda^2 (2t+1) / ((n-t+1)(m-t+1)) + 2\lambda p^t.$$

Hence, the probability above is correct to within 8.5×10^{-7}.

APPLICATION TO RNA EVOLUTION

Now we bring these ideas to bear on our RNA evolution problem. We have a set of 33 tRNA molecules and one 16**S** rRNA molecule from *E. coli*. In Bloch et al. (1983), matchings between 16**S** and each of the tRNAs were intensely studied. tRNA evolution is a complex topic and tRNA/tRNA comparisons were not made in this study. Table 4.1 shows the length of the longest exact match H_n between these sequences, along with estimates of significance or *p*-values $(1 - e^{-\lambda_n})$ from our Chen-Stein method. There are no exceptionally good matchings in this list, and so this analysis discounts any deep relationship between the sequences. In fact the *p*-values seem unusually large. In the 33 comparisons the minimum *p*-value is 0.26. Still we should not give up the search. One estimate puts the origin of these sequences at 3 billion years ago. We should not expect large segments of sequence to be preserved in every position over such vast amounts of time. Instead, mutations such as substitutions, insertions, and deletions will accumulate, greatly complicating our task. It is possible that the hy-pothesis of common origin is correct and that so much evolutionary change has taken place that no significant similarity remains. The next section, "Two Behaviors Suffice," examines the results of this search for unusual similarity using more subtle sequence comparison algorithms.

Table 4.1 Exact Match P-Values

tRNA	GenBank Locus	Length (n)	H_n	$1 - e^{-\lambda n}$	b_1
ala-1a	ECOTRA1A	76	9	0.26	1.87×10^{-5}
ala-1b	ECOTRA1B	76	9	0.26	1.87×10^{-5}
cys	ECOTRC	74	8	0.69	2.67×10^{-4}
asp-1	ECOTRD1	77	8	0.71	2.79×10^{-4}
glu-1	ECOTRE1	76	10	0.71	1.25×10^{-6}
glu-2	ECOTRE2	76	10	0.71	1.25×10^{-6}
phe	ECOTRF	76	9	0.26	1.87×10^{-5}
gly-1	ECOTRG1	74	7	0.99	3.90×10^{-3}
gly-2	ECOTRG2	75	6	1.00	5.70×10^{-2}
gly-3	ECOTRG3	76	9	0.26	1.87×10^{-5}
his-1	ECOTRH1	77	9	0.26	1.89×10^{-5}
ile-1	ECOTRI1	77	9	0.26	1.89×10^{-5}
ile-2	ECOTRI2	76	10	0.71	1.25×10^{-6}
lys	ECOTRK	76	6	1.00	5.78×10^{-2}
leu-1	ECOTRL1	87	8	0.76	3.19×10^{-4}
leu-2	ECOTRL2	87	8	0.76	3.19×10^{-4}
leu-5	ECOTRL5	87	9	0.29	2.16×10^{-5}
met-f	ECOTRMF	77	9	0.26	1.89×10^{-5}
met-m	ECOTRMM	77	8	0.71	2.79×10^{-4}
asn	ECOTRN	76	7	0.99	4.01×10^{-3}
gln-1	ECOTRQ1	75	8	0.70	2.71×10^{-4}
gln-2	ECOTRQ2	75	8	0.70	2.71×10^{-4}
arg-1	ECOTRR1	76	7	0.99	4.01×10^{-3}
arg-2	ECOTRR2	77	7	0.99	4.07×10^{-3}
ser-1	ECOTRS1	88	8	0.76	3.23×10^{-4}
ser-3	ECOTRS3	93	9	0.31	2.33×10^{-5}
thr-ggt	ECOTRTACU	76	7	0.99	4.01×10^{-3}
val-1	ECOTRV1	76	8	0.70	2.75×10^{-4}
val-2a	ECOTRV2A	77	8	0.71	2.79×10^{-4}
val-2b	ECOTRV2B	77	9	0.26	1.89×10^{-5}
trp	ECOTRW	76	7	0.99	4.01×10^{-3}
tyr-1	ECOTRY1	85	8	0.75	3.11×10^{-4}
tyr-2	ECOTRY2	85	8	0.75	3.11×10^{-4}

H_n, the length of the longest exact match between the listed tRNA molecule and a 16S rRNA molecule; $1 - e^{-\lambda n}$, the p-value (estimate of significance) for n^{th} tRNA molecule; b_1, column entry is the calculated bound on b_1.

TWO BEHAVIORS SUFFICE

In this section we describe a statistic that provides a link between the sections "Global Sequence Comparison" and "Local Sequence Comparison" of this chapter. This statistic is the score of the best matching intervals between two sequences, where nonidentities in the alignments receive penalties. In the section "Global Sequence Comparisons," we showed that the growth of score of global alignments of random sequences is linear with sequence length. In the section "Local Sequence Comparisons," we showed that the number of long runs of exact matches between random sequences has an approximate Poisson distribution. Below we show that the Poisson distribution implies that, for exact matching, the growth of longest run length is proportional to the logarithm of the product of sequence length. Then we state the result that all optimal alignments of a broad class have a score that has either logarithmic or linear growth, depending on the penalties for nonidentities. We will consider two sequences $\mathbf{A} = A_1 A_2 \ldots A_n$ and $\mathbf{B} = B_1 B_2 \ldots B_n$ of equal length n.

Recall that $p := \mathbf{P}(\text{two random letters are identical}) = \mathbf{P}(C_\alpha = 1)$. In the case of unknown alignments, $\lambda = \mathbf{E}(W)$ is given from equation (4.5) by

$$\lambda = p^t((n+n-2t+1)+(n-t)(n-t)(1-p)).$$

For $\lambda \approx 1$, we expect one run of length t. Then

$$\begin{aligned} 1 &= p^t((n+n-2t+1)+(n-t)(n-t)(1-p)) \\ &\approx p^t(nn(1-p)). \end{aligned}$$

Solving for t yields

$$t = \log_{1/p}(nn(1-p)).$$

Therefore the length of the longest run of identities grows like $\log_{1/p}(n^2) = 2\log_{1/p}(n)$.

To relax our stringent requirement of identities, we recall scoring for the alignments as introduced in the section "Global Sequence Comparisons." Extend the sequence $A_1A_2\ldots A_n$ to $A_1^*A_2^*\ldots A_L^*$ by inserting gaps "–" and similarly extend $B_1B_2\ldots B_n$ to $B_1^*B_2^*\ldots B_L^*$.

Define

$$S(\mathbf{A},\mathbf{B}) = \max\sum_{i=1}^{L} s(A_i^*,B_i^*),$$

where

$$s(A,B) = \begin{cases} +1 & \text{if } A=B \\ -\mu & \text{if } A\neq B \end{cases},$$

$$s(-,B) = s(A,-) = -\delta,$$

and $\mu \geq 0, \delta \geq 0$. The maximum is extended over all ways of inserting gaps and all L.

In Smith and Waterman (1981) and Waterman and Eggert (1987), dynamic programming algorithms are presented to compute

$$H(\mathbf{A},\mathbf{B}) = H(\mathbf{A},\mathbf{B};\mu,\delta) = \max\{S(I,J):\ I\subset\mathbf{A}, J\subset\mathbf{B}\}$$

in time $O(n^2)$. By $I\subset\mathbf{A}$, for example, we mean all $I = A_iA_{i+1}\ldots A_j$, where $1\leq i\leq j\leq n$. This algorithm was designed to study situations like our 16**S** rRNA/tRNA relationships. We are searching for segmental alignments that are not necessarily perfect matchings but are unusually good matches. After some discussion of the statistical properties of $H(\mathbf{A},\mathbf{B};\mu,\delta)$, we will apply the algorithm to our data.

The statistic $H(\mathbf{A},\mathbf{B};\mu,\delta)$ is for one of the so-called local alignment algorithms. However, when the penalties μ and δ are set to 0, the algorithm computes a global alignment. The results in the section "Global Sequence Comparisons" imply that

$$H(\mathbf{A},\mathbf{B};0,0) \sim a\cdot n;$$

but when the parameters are set to ∞, the results in the section "Local Sequence Comparisons" imply that

$$H(\mathbf{A},\mathbf{B};\infty,\infty) \sim b \log n .$$

It is natural to ask if there are other growth rates. The answer is presented in Waterman et al. (1987) and Arratia and Waterman (1994), where the following result is proved: Assume both sequences have equal lengths n. There is a continuous curve in the nonnegative (μ,δ) plane such that when (μ,δ) belongs to F_0, the same component as $(0,0)$, the growth of H is linear with sequence length. When (μ,δ) belongs to F_∞, the same component as (∞,∞), the growth is logarithmic with sequence length. In any curve crossing from F_0 to F_∞ there is a phase transition in growth of the score $H(\mu,\delta)$. This behavior is quite general, and in Arratia and Waterman (1994) it is shown to hold with very general penalties for scoring matches, mismatches, and indels. The behavior of $H(\mathbf{A},\mathbf{B};\mu,\delta)$ when (μ,δ) lies on the line between F_0 and F_∞ remains an open question.

RNA EVOLUTION REVISITED

How do the results in the previous section apply to our comparisons of 16**S** rRNA with tRNAs? As we have seen, the matchings of Bloch et al. (1983) were the result of applying a local algorithm, and so we will apply the local algorithm H to the data and study the distribution of scores. The first task is to compare the sequences using the statistic $H(\mathbf{A},\mathbf{B};\mu,\delta)$ with $\mu = 0.9$ and $\delta = 2.1$. These values have been used in several database searches, and the growth of scores from aligning random sequences lies in the logarithmic region. The results of the algorithm applied to our data can be found in Table 4.2. No closed-form Chen-Stein method has been arrived at for alignments with indels, so the results are presented in number of standard deviations ($\#\sigma$) above the mean value for comparing two random sequences of similar lengths. (See Waterman and Vingron (1994) for recent work on estimating statistical significance.) The estimated mean as a function of the tRNA length is

$$H(\mathbf{A},\mathbf{B};\mu = 0.9,\delta = 2.1) \ = \ 5.04 \log n - 30.95 ,$$

Table 4.2 Scores and Alignment Statistics

tRNA	Score	# σ	Matches	mms.	Indels
ala-1a	12.2	-.02	14	2	0
ala-1b	12.2	-0.1	14	2	0
cys	21.0	6.2	40	10	5
asp-1	10.8	-1.1	22	8	2
glu-1	10.9	-0.8	21	9	1
glu-2	12.8	0.6	22	8	1
phe	13.0	0.6	32	10	5
gly-1	9.4	-1.4	15	4	1
gly-2	9.5	-1.2	35	15	6
gly-3	14.4	1.5	41	14	7
his-1	13.2	1.1	28	12	2
ile-1	13.6	0.9	41	26	2
ile-2	14.0	1.3	35	10	6
lys	10.7	-0.5	23	7	3
leu-1	13.8	0.7	49	28	5
leu-2	11.7	-0.7	33	17	3
leu-5	13.4	0.4	36	14	5
met-f	12.0	-.03	44	20	7
met-m	11.4	-0.2	21	4	3
asn	15.3	2.4	33	13	3
gln-1	11.8	0.1	23	8	2
gln-2	12.1	0.2	26	11	2
arg-1	13.3	0.7	48	23	7
arg-2	12.8	0.3	26	8	3
ser-1	11.1	-1.3	29	11	4
ser-3	13.8	0.3	42	18	6
thr-ggt	10.1	-1.3	15	1	2
val-1	11.9	-0.2	22	9	1
val-2a	11.3	-0.7	14	3	0
val-2b	11.3	-0.4	14	3	0
trp	11.0	-0.7	22	10	1
tyr-1	11.7	-0.4	31	17	2
tyr-2	10.9	-0.9	42	19	7

σ, the number of standard deviations above the mean value (for comparing the two random sequences of similar lengths); mms., mismatches; indels, insertions/deletions.

while the standard deviation is estimated to be $\hat{\sigma} = 1.49$. In contrast to Table 4.1, there is one tRNA, that for cystine, that scores exceptionally high. The tails of the extremal distributions in the logarithmic region probably behave like $\exp(-\lambda t)$, where t is the test value as in the section "Local Sequence Comparisons" and λ is a constant, but this has not yet been proven rigorously. The usual intuition informed by the tail of a normal distribution has the probabilities behaving like $\exp(-t^2)/2$, which converges to 0 much more rapidly than the Poisson or exponential tails. Thus except for the cystine score, the remaining scores look very much like scores from random sequences. Simulations were performed, and the score 21.0 has an approximate p-value of 10^{-3}, so that it is not possible to dismiss this matching for statistical reasons alone. As far as we know, no one has offered a biological explanation of this interesting match. As to the hypothesis of Bloch et al. (1983), while their work concluded that "matches are too frequent and extensive to be attributed to coincidence," it is not supported by the data but is instead the result of incorrect estimation of p-values. This data set received their most extensive analysis, and they concluded that over 30 percent of the matchings between *E. coli* 16**S** rRNA and tRNAs were significant at the level $\alpha = 0.10$. Correct estimates show about 10 percent of the matching at the level $\alpha = 0.10$. While the origin of life may be hiding in tRNA and 16**S** rRNA, it remains elusive.

REFERENCES

Aldous, D.J., 1989, *Probability Approximations via the Poisson Clumping Heuristic*, New York: Springer-Verlag.

Arratia, R.A., L. Goldstein, and L. Gordon, 1989, "Two moments suffice for Poisson approximation: The Chen-Stein method," *Annals of Probability* **17**, 9-25.

Arratia, R.A., L. Goldstein, and L. Gordon, 1990, "Poisson approximation and the Chen-Stein method," *Statistical Science* **5**, 403-434.

Arratia, R.A., and M.S. Waterman, 1994, "A phase transition for the score in matching random sequences allowing deletions," *Annals of Applied Probability* **4**, 200-225.

Bloch, D.P., B. McArthur, R. Widdowson, D. Spector, R.C. Guimaraes, and J. Smith, 1983, "tRNA-rRNA sequence homologies: Evidence for a common evolutionary origin?," *Journal of Molecular Evolution* **19**, 420-428.

Chung, K.L., 1974, *A Course in Probability Theory,* 2nd ed, San Diego, CA: Academic Press.

Chvátal, V, and D. Sankoff, 1975, "Longest common subsequences of two random sequences," *Journal of Applied Probability* **12**, 306-315.

Deken, J., 1979, "Some limit results for longest common subsequences," *Discrete Mathematics* **26**, 17-31.

Erdös, P., and A. Rényi, 1970, "On a new law of large numbers," *Journal d' Analyse Mathematique* **22**, 103-111. Reprinted in 1976 in *Selected Papers of Alfred Rényi,* Vol. 3, 1962-1970, Budapest: Akadémiai Kiadó.

Feller, W., 1968, *An Introduction to Probability Theory and Its Applications,* Vol. 1, 3rd ed., New York: Wiley and Sons.

Goldstein, L., 1990, "Poisson approximation and DNA sequence matching," *Communications in Statistics. Part A: Theory and Methods* **19**, 4167-4179.

Kingman, J.F.C., 1973, "Subadditive ergodic theory," *Annals of Probability* **1**, 883-909.

Smith, T.F., and M.S. Waterman, 1981, "Identification of common molecular subsequences," *Journal of Molecular Biology* **147**, 195-197.

Steele, J.M., 1986, "An Efron-Stein inequality for nonsymmetric statistics," *Annals of Statistics* **14**, 753-758.

Waterman, M.S., and M. Eggert, 1987, "A new algorithm for best subsequence alignments with application to tRNA-rRNA comparisons," *Journal of Molecular Biology* **197**, 723-728.

Waterman, M.S., L. Gordon, and R. Arratia, 1987, "Phase transitions in sequence matches and nucleic acid structure," *Proceedings of the National Academy of Sciences USA* **84**, 1239-1243.

Waterman, M.S., and M. Vingron, 1994, "Rapid and accurate estimates of statistical significance for data base searches," *Proceedings of the National Academy of Sciences USA* **91**, 4625-4628.

Chapter 5
Calibrating the Clock: Using Stochastic Processes to Measure the Rate of Evolution

Simon Tavaré
University of Southern California

Deoxyribonucleic acid (DNA) sequences record the history of life. Although DNA replication is remarkably accurate, mutations do occur at a small but nonnegligible rate, with the result that an individual's descendants begin to diverge in DNA sequence over time. By examining DNA sequences among different species or among different individuals within a single species, it is possible to reconstruct aspects of their evolutionary history. Such studies have been pursued with special interest in the human, where an unusual DNA sequence called the mitochondrial genome has been used to trace human migrations and human evolution. The author shows how mathematical tools from the theory of stochastic processes assist in calibrating the molecular clock inherent in DNA sequences.

While DNA sequences are transmitted from parent to child with remarkable fidelity, mutations do occur at a small but nonnegligible rate, with the result that an individual's descendants begin to diverge in DNA sequence over time. Some mutations are deleterious and are eliminated by natural selection, but many are thought to be selectively neutral and thus accumulate at a roughly steady rate—providing a molecular clock for measuring the time since two species or two individuals within a species shared a common ancestor. In this manner, it is possible to reconstruct an evolutionary tree and even estimate the times of key separation events.

Different biological sequences within an organism may obey different clocks. The amino acid sequence of a protein encoded by a gene changes more slowly than the DNA sequence of the underlying gene because many amino acid changes may be selectively disadvantageous (because they disrupt function). On the other hand, a significant proportion of DNA changes may be selectively neutral because they create a synonymous codon (that is, one that specifies the same amino acid). Similarly, DNA regions within genes change at a slower rate than the DNA sequences located between genes. Accordingly, evolutionary studies of distant species are often carried out by examining amino acid sequences of proteins, while evolutionary comparisons among more closely related species are better done by examining DNA sequences within or between genes.

To study evolution within a single species such as the human, it is often useful to study DNA sequences that change at especially rapid rates. The mitochondrial genome provides an ideal substrate for such studies. The mitochondrion is an organelle found in the cytoplasm of eukaryotic cells, whose primary role is to generate high-energy compounds that the cell uses to drive chemical reactions. Although the mitochondria use many proteins that are encoded by genes in the cell nucleus, each mitochondrion has its own small circular chromosome that encodes a few dozen genes essential for its function.

In the human, the mitochondrial genome consists of 16,569 base pairs whose DNA sequence has been completely determined (Anderson et al., 1981). Human mitochondria are inherited only from the mother, and so their genealogy is considerably simpler to follow than for genes encoded in the nucleus (which are inherited from both parents and are subject to recombination between the two homologous copies in the cell). Conveniently for evolutionary studies, mitochondrial DNA (mtDNA) has an increased rate of nucleotide substitution compared to nuclear genes, owing to the presumed absence of certain DNA repair mechanisms. Moreover, the mitochondrial genome contains certain regions that are particularly tolerant of mutation, that is, appear to be subject to little selective pressure (Avise, 1986) and thus show a great deal of variation. In all, the mitochondrial genome may be evolving 10 times faster than the nuclear genome.

For these reasons, molecular population geneticists have carried out many studies of the DNA sequences of mitochondrial variable regions in many human populations (Di Rienzo and Wilson, 1991; Horai and

Hayasaka, 1990; Vigilant et al., 1989, 1991; Ward et al., 1991). Studies of mitochondrial sequences of different Native American tribes strongly suggest that there were multiple waves of colonization of North America by migrant groups from Asia, and even allow one to estimate the dates of these events (Schurr et al., 1990; Ward et al., 1991). Assuming a constant evolutionary rate, the pattern of mutations between diverse human groups has been used to argue (Cann et al., 1987) that the mitochondria of all living humans descended from a mother that lived in Africa some 200,000 years ago—the so-called Eve hypothesis. Although the precise details of the hypothesis are disputed (Maddison, 1991; Nei, 1992; Templeton, 1992), the general power of the methodology is well accepted. (As an aside, the reader should note that the existence of a common ancestor—Eve, so to speak—is a mathematical necessity in any branching process that satisfies very weak conditions. The biological controversies pertain to when and where Eve lived.)

Each of these applications requires a knowledge of the rate at which mutations occur in an mtDNA sequence. Estimates of this rate have been obtained by comparing a single DNA sequence from each of several species whose times of divergence are presumed known. Divergence is calculated from the number of nucleotide differences between species (using methods that correct for the possibility of multiple mutations at a site), and rate estimates are obtained by dividing the amount of sequence divergence by the divergence time. For data taken from multiple individuals in a single population, one requires a model that takes account of the population genetic aspect of the sampling: individuals in the sample are correlated by their common ancestry. In this chapter, we describe the underlying stochastic structure of this ancestry and use the results to estimate substitution rates.

We have chosen to focus on rate estimation to give the chapter a single theme. We are not interested per se in statistical aspects of tests for selective neutrality of DNA differences; rather, we assume neutrality for the data sets discussed as examples. The techniques described here should be regarded as illustrative of the theoretical and practical problems that arise in sequence analysis of samples from closely related individuals. The emphasis is on exploratory methods that might be used to summarize the structure of such samples.

OVERVIEW

To illustrate the methods, we use a set of North American Indian mitochondrial sequences described in Ward et al. (1991). These authors sequenced the first 360 base pairs of the mitochondrial control region for a sample of 63 Nuu-Chah-Nulth (Nootka) Indians from Vancouver Island. The sample comprises individuals who were maternally unrelated for four generations, chosen from 13 of the 14 tribal bands. As a consequence the sample deviates from a truly random sample, although it will be treated as such for the purposes of this chapter. An important parameter in the analysis is the effective population size of the group. This is approximated by the number of reproducing females, giving a value of about 600 for the long-term effective population size N.

The most common DNA changes seen in mitochondria are transitions (changes from one pyrimidine base to the other or one purine base to the other, that is, `C` $\leftrightarrow$ `T` or `A` $\leftrightarrow$ `G`) rather than transversions (changes from a pyrimidine to a purine or vice versa). Indeed, the sequenced region shows no transversions, so that each site in the sequences has one of just two possible nucleotides. We focus on the pyrimidine (`C` or `T`) sites in the region. There are 201 such sites, in which 21 variable (or segregating) sites define 24 distinct sequences (called alleles or lineages). The details of the data, including the allele frequencies, are given in Table 5.1.

The parameter of particular interest here is θ, the population geneticist's stock in trade. The variable θ is a measure of the mutation rate in the region, and it figures in many important theoretical formulas in population genetics. For mitochondrial data, it is defined by

$$\theta = 2Nu,$$

where N is the effective population size referred to earlier, and u is the mutation rate per gene per generation. Once θ is estimated, we can estimate u if N is known or N if u is known. In what follows, we estimate the compound parameter θ rather than its components.

In the section immediately following, we begin by outlining the structure of the coalescent, a robust description of the genealogy of samples taken from large populations. The effects of mutation are superimposed on this genealogy in several ways. The classical case, which

Table 5.1 Nucleotide Position in Control Region

Position	69	88	91	124	149	162	166	194	200	219	233	247	255	267	271	275	301	302	304	319	339	Allele Frequencies
Site	1	2	3	4	5	6	7	8	9	10	11	12	13	14	15	16	17	18	19	20	21	
ID ref	T	C	C	C	T	C	T	T	C	C	C	C	C	C	C	T	T	T	C	T	T	
1	.	.	.	.	.	.	C	.	T	.	.	.	T	.	.	.	.	.	.	.	.	
2&3	.	.	.	.	.	.	.	.	T	.	.	.	T	.	.	.	.	.	.	.	.	3
4	.	.	.	.	.	.	.	.	T	.	.	.	T	.	.	.	.	.	.	.	C	
5	.	T	.	.	.	T	.	.	T	.	.	.	.	T	.	.	.	.	.	.	C	3
6	.	T	.	.	.	.	.	.	T	.	.	.	.	.	.	.	.	.	.	.	C	2
7	C	T	.	.	.	.	.	.	T	.	.	T	.	.	.	.	.	.	.	.	C	1
8,10&11	.	T	.	.	.	.	.	.	T	.	.	.	.	T	.	.	.	.	.	.	C	8
9	C	T	.	.	.	.	.	.	T	.	.	.	.	T	.	.	.	.	.	.	C	2
12&13	.	T	.	.	.	.	.	.	.	.	.	.	.	T	.	.	.	.	.	.	C	10
14	.	T	.	.	.	.	.	.	T	.	.	.	T	T	.	.	.	.	.	.	C	1
15	.	T	.	.	.	.	.	.	T	.	.	.	T	T	.	.	C	.	.	.	C	2
16	.	.	.	.	.	.	.	.	T	T	.	.	.	.	.	.	.	.	T	.	C	1
17	.	.	.	T	.	.	.	.	T	.	.	.	.	.	.	.	C	.	.	.	C	1
18	.	.	.	T	.	.	.	.	T	.	.	.	.	.	.	.	.	C	.	.	C	2
19	.	.	T	.	C	.	.	.	T	.	.	.	.	.	T	.	.	C	.	.	C	1
20	.	.	.	.	.	.	.	.	T	.	.	.	.	.	.	.	.	C	.	.	C	3
21	.	.	.	.	.	.	.	.	T	.	.	.	.	.	.	.	.	C	.	.	C	3
22	C	.	.	.	.	.	.	.	T	.	.	.	.	.	.	.	.	C	.	.	.	3
23	.	.	.	.	.	.	.	.	T	T	.	.	.	.	.	C	.	C	.	.	.	1
24	.	.	.	.	.	.	.	.	T	.	.	.	.	.	.	C	.	C	T	.	.	7
25	.	.	.	.	.	.	.	.	T	T	.	.	.	.	.	C	.	C	T	C	.	3
26	.	.	.	.	.	.	.	.	.	T	.	.	.	.	.	C	.	C	T	C	.	1
27	.	.	.	.	.	.	C	C	.	.	.	.	.	.	.	.	.	.	.	.	.	1
28	.	.	.	.	.	.	C	C	.	.	T	.	.	.	.	.	.	.	.	.	.	1

NOTE: These mitochondrial data from Ward et al. (1991, Figure 1) are the variable pyrimidine positions in the control region. Position 69 corresponds to position 16,092 in the human reference sequence published by Anderson et al. (1981). The ID numbers correspond to those given in Ward et al. (1991, Figure 1).

records the allelic partition of the sample, leads to the sampling theory of the infinitely-many-alleles model initiated by Ewens (1972). The Ewens sampling formula is then described, followed by a brief digression into the simulation structure of mutations in the coalescent, both in top-down and bottom-up form. Next, the infinitely-many-sites model is introduced as a simple description of the detailed structure of the segregating sites in the sample. Finally, we return to classical population genetics theory, albeit from a coalescent point of view, to discuss the structure of K-allele models. This in turn develops into the study of the finitely-many-sites models, which play a crucial role in the study of sequence variability when back substitutions are prevalent.

In the next section we digress to present a mathematical vignette in the area of random combinatorial structures. The Ewens sampling formula was derived as a means to analyze allozyme frequency data that became prevalent in the late 1960s. Current population genetic data is more sequence oriented and requires more detailed models for its analysis. Nonetheless, the combinatorial structure of the Ewens sampling formula has recently emerged as a useful approximation to the component counting process of a wide range of combinatorial objects, among them random permutations, random mapping functions, and factorization of polynomials over a finite field. We show how a result of central importance in the development of statistical inference for molecular data has a new lease on life in an area of discrete mathematics.

The final section briefly discusses some of the outstanding problems in the area, with particular emphasis on likelihood methods for coalescent processes. Some aspects of the mathematical theory, for example, measure-valued diffusions, are also mentioned, together with applications to other, more complicated, genetic mechanisms.

THE COALESCENT AND MUTATION

The genealogy of a sample of n genes (that is, stretches of DNA sequence) drawn at random from a large population of approximately constant size may be described in terms of independent exponential random variables $T_n, T_{n-1}, \ldots, T_2$ as follows. The time T_n during which the sample has n distinct ancestors has an exponential distribution with parameter $n(n-1)/2$, at which time two of the lines are chosen at random to coalesce, giving the sample $n-1$ distinct ancestors. The time T_{n-1} during which the sample has $n-1$ such ancestors is exponentially distributed with parameter $(n-1)(n-2)/2$, at which point two more ancestors are chosen at random to coalesce. This process of coalescing continues until the sample has two distinct ancestors. From that point, it takes an exponential amount of time T_2 with parameter 1, to trace back to the sample's common ancestor. For our purposes, the time scale is measured in units of *N* generations, where *N* is the (effective) size of the population from which the sample was drawn. This structure, made explicit by Kingman (1982a,b), arises as an approximation for large *N* to many models of reproduction, among them the Wright-Fisher and Moran models. A sample path of a coalescent with $n = 5$ is shown in Figure 5.1.

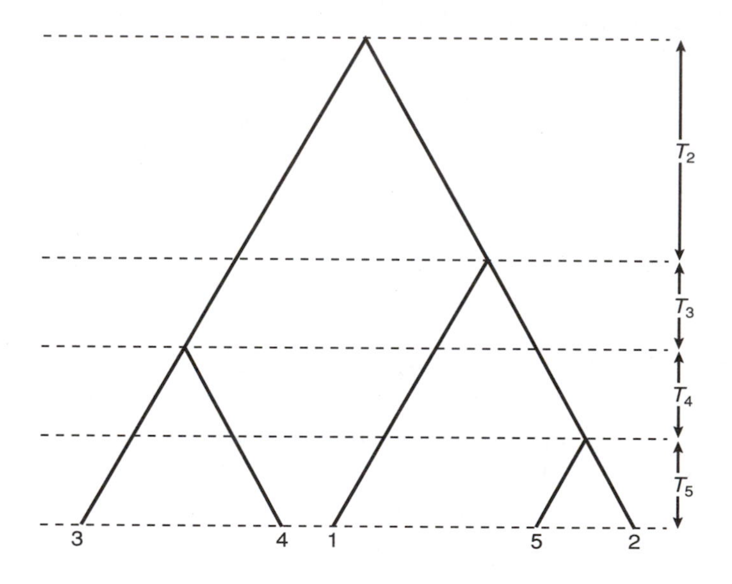

FIGURE 5.1 Sample path of the coalescent for $n = 5$. T_j denotes the time during which the sample has j distinct ancestors. T_j has an exponential distribution with mean $2/j(j-1)$.

From the description of the genealogy, it is clear that the time τ_n back to the common ancestor has mean

$$E(\tau_n) = \sum_{j=2}^{n} E(T_j) = \sum_{j=2}^{n} \frac{2}{j(j-1)} = 2\left(1 - \frac{1}{n}\right),$$

or approximately $2N$ generations for large sample sizes. Further aspects of the structure of the ancestral process may be found in Tavaré (1984). Rather than focus further on such issues, we describe how the genealogy may be used to study the genetic composition of the sample.

To this end, assume that in the population from which the sample was drawn there is a probability u that any gene mutates in a given generation, mutation acting independently for different individuals. In looking back r generations through the ancestry of a randomly chosen gene, the number of mutations along that line is a binomial random variable with parameters r and u. If we measure time in units of N generations, so that $r = \lfloor Nt \rfloor$ (that is, r is Nt rounded down to the next lower integer), and assume that $2Nu \to \theta$ as $N \to \infty$, then the Poisson approximation to the binomial distribution shows that the number of mutations in time t has in the limit a Poisson distribution with mean $\theta t / 2$. This argument can be extended to show that the mutations that arise on different branches of the coalescent tree follow independent Poisson processes, each of rate $\theta / 2$. For example, the total number of mutations μ_n that occur in the history of our sample back to its common ancestor has a mixed Poisson distribution—given $T_n, T_{n-1}, \ldots, T_2$, μ_n has a Poisson distribution with mean $\frac{1}{2}\theta \sum_{j=2}^{n} jT_j$. The mean and variance of the number of mutations are given by Watterson (1975):

$$E(\mu_n) = \frac{\theta}{2} \sum_{j=2}^{n} jE(T_j) = \theta \sum_{j=1}^{n-1} \frac{1}{j}, \tag{5.1}$$

and

$$\mathrm{Var}(\mu_n) = \theta \sum_{j=1}^{n-1} \frac{1}{j} + \theta^2 \sum_{j=1}^{n-1} \frac{1}{j^2}. \tag{5.2}$$

We are now in a position to describe the effect that mutation has on the individuals in the sample.

The Ewens Sampling Formula

Motivated by the realization that mutations in DNA sequences could lead to an essentially infinite number of alleles at the given locus, Kimura and Crow (1964) advocated modeling the effects of mutation as an infinitely-many-alleles model. In this process, a gene inherits the type of its ancestor if no mutation occurs and inherits a type not currently (or previously) existing in the population if a mutation does occur. In such a process the genes in the sample are thought of as unlabeled, so that the experimenter knows whether two genes are different, but records nothing further about the identity of alleles. In this case the natural statistic to record about the sample is its configuration $\mathbf{C}_n \equiv (C_1, C_2, \ldots, C_n)$, where

$$C_j = \text{number of alleles represented } j \text{ times.}$$

Of course, $C_1 + 2C_2 + \ldots + nC_n = n$, and the number of alleles in the sample is

$$K_n \equiv C_1 + C_2 + \ldots + C_n . \tag{5.3}$$

The sampling distribution of $\mathbf{C}_n$ was found by Ewens (1972):

$$\mathbf{P}(\mathbf{C}_n = \mathbf{a}) = \frac{n!}{\theta_{(n)}} \prod_{j=1}^{n} \left(\frac{\theta}{j}\right)^{a_j} \frac{1}{a_j!}, \tag{5.4}$$

for $\mathbf{a} = (a_1, a_2, \ldots, a_n)$ satisfying $a_j \geq 0$ for $j = 1, 2, \ldots, n$ and $\sum_{j=1}^{n} ja_j = n$, and where

$$\theta_{(n)} \equiv \theta(\theta + 1) \cdots (\theta + n - 1) .$$

From (5.4) it follows that

$$\mathbf{P}(K_n = k) = \frac{|S_n^k|\,\theta^k}{\theta_{(n)}}, \quad k = 1,2,\ldots,n, \tag{5.5}$$

and

$$\mathbf{E}(K_n) = \sum_{j=0}^{n-1} \frac{\theta}{\theta + j}, \tag{5.6}$$

S_n^k being the Stirling number of the first kind. From (5.5) and (5.4) it follows that K_n is sufficient for θ, so that the information in the sample relevant for estimating θ is contained just in K_n. This allows us (Ewens, 1972, 1979) to calculate the maximum likelihood (and moment) estimator of θ as the solution $\hat{\theta}$ of the equation

$$k = \sum_{j=0}^{n-1} \frac{\hat{\theta}}{\hat{\theta} + j}, \tag{5.7}$$

where k is the number of alleles observed in the sample. In large samples, the estimator $\hat{\theta}$ has variance given approximately by

$$\mathrm{Var}(\hat{\theta}) \approx \theta \left(\sum_{k=2}^{n} \frac{k-1}{(\theta + k - 1)^2} \right)^{-1}. \tag{5.8}$$

For the pyrimidine sequence data described above in the "Overview" section, there are $k = 24$ alleles. Solving equation (5.7) for $\hat{\theta}$ gives $\hat{\theta} = 10.62$, with a variance of 9.89. An approximate 95 percent confidence interval for θ is therefore 10.62 ± 6.29. This example serves to underline the variability inherent in estimating θ from this model. The pyrimidine region comprises 201 sites, so that the per site substitution rate is estimated to be 0.053 ± 0.031.

The goodness of fit of the model to the data may be assessed by using the sufficiency of K_n for θ: given K_n, the conditional distribution of the allele frequencies is independent of θ. Ewens (1972, 1979) gives further details on this point. To describe alternative goodness-of-fit methods, we return briefly to the probabilistic structure of mutation in the coalescent.

Forwards and Backwards in the Tree

Hudson (1991) describes many situations in which simulation of genealogical trees is useful. In its simplest form, the idea is to construct (a simulation of) a coalescent tree, with times and branching order, and then superimpose the effects of mutation on this tree using the Poisson nature of the mutation process. In this section we make use of two equivalent descriptions of the effects of mutation in the coalescent tree in which the mutation and coalescence events evolve simultaneously.

Top-down

The first of these methods is a very useful "top-down" scheme exploited by Ethier and Griffiths (1987) in the context of the infinitely-many-sites model. We start at the common ancestor of the sample and think of the genetic process running down to the sample. Just after the first split, we have a sample of two individuals, each of identical genetic type. Attach to each individual a pair of independent exponential alarm clocks—one of rate $\theta/2$, the second of rate 1/2—and suppose the clocks are independent for different individuals. The θ clocks will determine mutations, the other clocks split times. Now watch the clocks until the first one rings: if a θ-clock rings, a mutation occurs in that gene, whereas if one of the other clocks rings, a split occurs in which that gene is copied, now making a sample of three individuals. Using the standard "competing exponentials" argument, the probability that a mutation occurs first is

$$\frac{\theta/2+\theta/2}{\theta/2+\theta/2+1/2+1/2}=\frac{\theta}{\theta+1},$$

whereas a split occurs first with probability $1/(\theta+1)$. Furthermore, given that a mutation occurs first, the gene in which it occurs is chosen uniformly and at random, and given that a split occurs first, the gene that is copied is chosen uniformly and at random.

Once an event occurs, the process repeats itself in a similar way. Suppose, then, that there are currently m genes in the sample. Attach independent mutation clocks of rate $\theta/2$ and independent split clocks of rate $(m-1)/2$ to each of the m genes and wait for one to ring. The probability that a mutation clock rings first is $\theta/(\theta+m-1)$, and, given that a mutation occurs first, the gene that mutates is chosen uniformly and at random. Similarly, the probability that a split occurs first is $(m-1)/(\theta+m-1)$, with the splitting gene being chosen at random from the m possibilities.

The only wrinkle left is to describe the rule that tells us when to stop generating splits or mutations. In order to have the right distribution for the numbers of mutations when the sample has n ancestors, we must run until the first split after n, discard the last observation, and then stop.

This simple scheme can be used effectively to simulate observations from extremely complex mutation mechanisms using only Bernoulli random variables, and provides a way of generating and storing the effects of each of the mutations. Some examples are given in the following sections.

Bottom-up

The second scheme, which proves very useful for deriving recurrence relations for the distribution of allele configurations, is the "bottom-up" method. In this case, the idea is to use the exponential alarm clocks from the bottom of the tree (that is, beginning at the sample) and run up to the common ancestor at the top. If we look up from the sample of size n toward the root, the probability that we will encounter a mutation before a coalescence is $\theta/(\theta+n-1)$, and the probability that a coalescence will occur first is $(n-1)/(\theta+n-1)$. The probability distribution of the configuration at the tips may then be related to the distribution of the configuration at the mutation or coalescence time.

To illustrate how this works, consider the infinitely-many-alleles mutation structure. Suppose that the current configuration consists of counts $\mathbf{a}=(a_1,a_2,\ldots,a_n)$ with $a_n=0$, and let $\mathbf{P}_n(\mathbf{a})$ denote the probability of this configuration. If the first event in the past is a coalescence, then the configuration of $n-1$ genes must have been

$$(a_1,\ldots,a_{j-1},a_j+1,a_{j+1}-1,a_{j+2},\ldots,a_{n-1})$$

for some $j = 1, 2, \ldots, n-2$, and a gene in class *j* must be chosen to have an offspring. Since this last event has probability $j(a_j + 1)/(n-1)$, the contribution to $\mathbf{P}_n(\mathbf{a})$ from such terms is

$$\frac{n-1}{\theta + n - 1}\left(\sum_{j=1}^{n-2} \frac{j(a_j+1)}{n-1} \mathbf{P}_{n-1}(a_1, \ldots a_{j-1}, a_j + 1, a_{j+1} - 1, a_{j+2}, \ldots, a_{n-1})\right) \tag{5.9}$$

If, on the other hand, the first event in the past was a mutation, then the configuration must have been either

$$(a_1 - 1, a_2, \ldots, a_{j-2}, a_{j-1} - 1, a_j + 1, a_{j+1}, \ldots, a_{n-1}, 0)$$

and the mutation occurred to a gene in a *j* class, $j = 3, 4, \ldots, n-1$ (probability $j(a_j + 1)/n$), or

$$(a_1 - 2, a_2 + 1, a_3, \ldots, a_{n-1}, 0)$$

and the mutation occurred to a gene in the 2 class (probability $2(a_2 + 1)/n$), or

$$(a_1, \ldots, a_{n-1}, 0)$$

and the mutation occurred to a singleton gene (probability a_1/n). Finally, the configuration could have been

$$(a_1 - 1, a_2, \ldots, a_{n-2}, a_{n-1} - 1, 1)$$

and the mutation occurred in the *n* class (probability 1). Combining all these possibilities and adding the term in (5.9) gives

$$\mathbf{P}_n(\mathbf{a}) = \frac{\theta}{\theta+n-1}(\mathbf{P}_n(a_1-1,a_2,\ldots,a_{n-2},a_{n-1}-1,1)$$
$$+\frac{a_1}{n}\mathbf{P}_n(a_1,\ldots,a_{n-1},0)$$
$$+\frac{2(a_2+1)}{n}\mathbf{P}_n(a_1-2,a_2+1,a_3,\ldots,a_{n-1},0)$$
$$+\sum_{j=3}^{n-1}\frac{j(a_j+1)}{n}\mathbf{P}_n(a_1-1,a_2,\ldots,a_{j-2},a_{j-1}-1,a_j+1,a_{j+1},\ldots,a_{n-1},0)$$
$$+\frac{n-1}{\theta+n-1}(\sum_{j=1}^{n-2}\frac{j(a_j+1)}{n-1}\mathbf{P}_{n-1}(a_1,\ldots,a_{j-1},a_j+1,a_{j+1}-1,a_{j+2},\ldots,a_{n-1})). \quad (5.10)$$

The only case not covered by equation (5.10) is the one in which $\mathbf{a} = (0,\ldots,0,1)$. In this case the previous event had to be a coalescence, and so

$$\mathbf{P}_n(0,\ldots,0,1) = \frac{n-1}{\theta+n-1}\mathbf{P}_{n-1}(0,\ldots,0,1). \quad (5.11)$$

The persistent reader will be able to verify that $\mathbf{P}_n(\mathbf{a})$ given by the Ewens sampling formula (5.4) does indeed satisfy equations (5.10) and (5.11).

The Infinitely-Many-Sites Model

The infinitely-many-sites model of Kimura (1969) and Watterson (1975) is the simplest description of the evolution of a population of DNA sequences. The sites in the sequences are completely linked, and each mutation that occurs in the ancestral tree of the sample introduces a new segregating site into the sample. In this process, each new mutation occurs at a site not previously segregating—new mutations arise just once. It follows that at each segregating site, the sample may be classified as type 0 (ancestral) or type 1 (mutant). Of course, in practice we do not know which is which. The sequences in the sample may now be described by strings of 0s and 1s. If distinct sequences are treated as alleles, then the sampling theory is reduced to that covered by the Ewens sampling formula.

The number S_n of segregating sites is an important summary statistic for the sample. Since each new mutation produces a segregating site, it follows that $S_n = \mu_n$, the number of mutations in the ancestral tree. The mean and variance of S_n are therefore given by (5.1) and (5.2), respectively.

The number of segregating sites has been studied extensively for many variants of the infinitely-many-sites process, including the effects of selection and recombination, for example. Hudson (1991) gives an accessible summary of this work. When there is no recombination, the fundamental results have been established by Watterson (1975), Ethier and Griffiths (1987), and Griffiths (1989).

Watterson (1975) parlayed the moments of S_n into an unbiased estimator $\tilde{\theta}$ of θ, namely,

$$\tilde{\theta} = \frac{S_n}{H_n}, \tag{5.12}$$

with variance

$$\mathrm{Var}(\tilde{\theta}) = \frac{\mathrm{Var}(S_n)}{H_n^2}.$$

where $H_n = \sum_1^{n-1} \frac{1}{j}$. Note that $\tilde{\theta}$ does not depend on knowing which type at a site is ancestral and does not make full use of the data. For the pyrimidine data, there are 21 segregating sites, giving an approximate 95 percent confidence interval for θ of 4.46 ± 3.10. This should be compared to the estimate of 10.62 ± 6.29 obtained from the Ewens sampling formula.

Now think of the data as an $n \times s$ matrix of 0s and 1s, s being the number of segregating sites in the sample. When 0 is known to be ancestral in each site, Griffiths (1987) established that the data are consistent with the infinitely-many-sites model as long as in any set of three rows of the matrix, at most one of the patterns

$$\begin{pmatrix}0\\1\\1\end{pmatrix}, \begin{pmatrix}1\\0\\1\end{pmatrix}, \begin{pmatrix}1\\1\\0\end{pmatrix}$$

occurs. This is equivalent to the pairwise compatibility condition for binary characters established by Estabrook et al. (1976) and McMorris (1977): two sites are compatible if two or fewer of the patterns 01, 10, 11 occur. When the ancestral state is unknown, an analogous result holds: two sites are compatible if at most three of the patterns 00, 01, 10, 11 occur.

This translates into a simple test of whether a given set of binary site data is consistent with the infinitely-many-sites model. If in all pairs of columns at most three of the patterns 00, 01, 10, 11 occur, then there is at least one labeling of the sites that is consistent. McMorris (1977) proved that consistent data remain consistent when the most frequent type is taken as ancestral.

In practice, back mutations and recombination make most molecular data inconsistent with this model. However, it is worthwhile to look for maximal subsets of sites that are consistent, as this provides a way to identify regions of the sequence with simple structure. For the pyrimidine data described in Table 5.1, the maximal consistent set has 14 sites, those in positions 2-8, 11-12, 14-16, and 20-21. The remaining 7 sites have some inconsistencies, attributable to back substitutions, for example.

Of the $2^{14} = 16{,}384$ possible relabelings of the consistent set, just 16 are consistent. Each of these labelings is associated with a genealogical tree that describes the relationships between the mutations in the coalescent. The precise definition of the (equivalence class of) trees is given in Ethier and Griffiths (1987) and Griffiths (1989). The tree is equivalent to those built using compatibility methods for binary characters; see Felsenstein (1982, pp. 389-393) for a detailed discussion and references. The nodes in the tree represent the mutations that have generated the segregating sites, and the tips represent the sequences. A convenient algorithm for finding these trees is provided by Griffiths (1987), who also shows (Griffiths, 1989) how the probability of a tree with a given ancestral labeling can be computed under the infinitely-many-sites model. Griffiths' program PTREE can then be used to construct true maximum likelihood estimators of the parameter θ. It can also be used to compare "likelihoods" of the different ancestral labelings. The corresponding theory of the unrooted genealogical trees that arise when the ancestral labeling is unknown has recently been developed by Griffiths and Tavaré (1994c), and this leads to a practical computational method for estimating θ by maximum likelihood.

Our analysis of the mitochondrial data set has shown that while parts of the region are consistent with a simple evolutionary model, there are sites which are behaving in a more complicated way. In the next section, we describe a finitely-many-sites model that is useful for modeling regions in which back mutations have occurred.

K-Allele Models

We turn first to the *K*-allele model. In this process, we assume that there are *K* possible alleles at the locus in question. When a mutation occurs to an allele of type *i*, there is a probability m_{ij} that the resulting allele is of type *j*. To allow for different rates of substitution for different alleles, we can have $m_{ii} > 0$, and we write $M = (m_{ij})$. The effects of mutation along a given line are now modeled by a continuous time Markov chain whose transition matrix $P(t) \equiv (p_{ij}(t))$ gives the probabilities that a gene of type *i* has been replaced by a descendant gene of type *j* a time *t* later. Indeed,

$$P(t) = \exp\left(\frac{\theta t}{2}(M - I)\right),$$

where *I* is the $K \times K$ identity matrix, so that the generator of the mutation process is

$$Q \equiv (q_{ij}) = \frac{\theta}{2}(M - I). \tag{5.13}$$

It is worth pointing out that a given Q matrix can be represented in more than one way in the form (5.13), so that θ, for example, is not identifiable without further assumptions. However, the rates q_{ij}, $j \neq i$ are identifiable. If Q has a stationary distribution $\pi = (\pi_1, \pi_2, \ldots, \pi_K)$ satisfying $\pi Q = 0$, $\sum_{j=1}^{K} \pi_j = 1$, and if the common ancestor of the sample has distribution π, then the distribution of a gene at any point in the tree is also π, and the process is stationary.

From the data analyst's perspective, the sample of n genes can be sorted into a vector $\mathbf{N} \equiv (N_1, N_2, \ldots, N_K)$ of counts, there being N_j alleles of type j in the sample. Surprisingly, the stationary distribution of $\mathbf{N}$ is known explicitly only for the special case

$$q_{ij} = \frac{1}{2}\varepsilon_j > 0, \quad j \neq i.$$

This is equivalent to the independent mutations case in which $\theta = \varepsilon_1 + \varepsilon_2 + \ldots + \varepsilon_K$, $\pi_i = \varepsilon_i / \theta$, and $m_{ij} = \pi_j$ for all i and j. In this case, Wright's Formula (Wright, 1968) can be used to show that

$$\mathbf{P}(\mathbf{N} = \mathbf{n}) = \binom{\theta + n - 1}{n}^{-1} \prod_{i=1}^{K} \binom{n_i + \theta\pi_i - 1}{n_i} \tag{5.14}$$

for

$\mathbf{n} = (n_1, n_2, \ldots, n_K)$, $n_j \geq 0$ for $j = 1, 2, \ldots, K$, and $n = n_1 + n_2 + \ldots + n_K$.

In the next section, we use this result for the case $K = 2$. If

$$Q = \frac{1}{2}\begin{pmatrix} -\alpha & \alpha \\ \beta & -\beta \end{pmatrix}, \tag{5.15}$$

then equation (5.14) specializes to

$$g(l) \equiv \mathbf{P}(N_1 = l) = \binom{\alpha + \beta + n - 1}{n}^{-1} \binom{l + \beta - 1}{l} \binom{n - l + \alpha - 1}{n - l} \tag{5.16}$$

for $l = 0, 1, \ldots, n$.

Because the sampling formula for general Q is not known explicitly, it is useful to have a way to compute it. Perhaps the simplest is an application of the bottom-up method described above. Define $q(\mathbf{n}) = \mathbf{P}(\mathbf{N} = \mathbf{n})$, and set $\mathbf{e}_j = (0,\ldots,0,1,0,\ldots,0)$, the 1 occurring in position j. Look up the tree to the first event that occurred. This is either a mutation (with probability $\theta / (\theta + n - 1)$) or a coalescence (with probability $(n-1)/(\theta + n - 1)$). By considering the configuration of the sample at this event, we see that $q(\mathbf{n})$ satisfies the recursion

$$\begin{aligned} q(\mathbf{n}) &= \frac{\theta}{\theta + n - 1}\Big(\sum_i \frac{n_i}{n} m_{ii} q(\mathbf{n}) + \sum_i \sum_{j \neq i} \frac{n_i + 1}{n} m_{ij} q(\mathbf{n} + \mathbf{e}_i - \mathbf{e}_j)\Big) \\ &\quad + \frac{n-1}{\theta + n - 1} \sum_j \frac{n_j - 1}{n - 1} q(\mathbf{n} - \mathbf{e}_j), \end{aligned} \tag{5.17}$$

where $q(\mathbf{n}) = 0$ if any $n_i < 0$, and $q(\mathbf{e}_i) = \pi_i^*$. The process is stationary if $\pi_i^* = \pi_i$ for all i. We exploit this recursion more fully in the section below on likelihood methods.

The Finitely-Many-Sites Models

We now have the machinery necessary to describe the finitely-many-sites model for molecular sequence data involving n sequences, each of s sites. The sites are thought of as completely linked, and each site is typically one of either 2 or 4 possibilities. At its grossest level, the finitely-many-sites model is "just" a K-allele model in which $K = 2^s$ or 4^s. From an inference point of view, however, there are far too many parameters in such a model, and some simplification is required. The simplest null model of sequence evolution is the case in which mutations occur at a rate of $\theta / 2$ per gene, but when a mutation occurs, a site is chosen at random to mutate and the base at that site changes according to a mutation matrix M. A slightly more general model might allow site j to mutate with probability $\mathbf{p}_j$, once more according to M. For a two-type classification of each site, the first model has two parameters to be estimated, and the second has $s + 1$. These schemes can be modified to allow for other correlation structures between the sites at the expense of more complicated methods of analysis.

Motivated by our sequence data, we concentrate on the two-state case and discuss methods for estimating the parameters of the simplest null model. At a single site, the model behaves exactly like a 2-allele process with

$$Q = \frac{\theta}{2s}(M - I),$$

because the per site substitution rate is θ / s. This has the structure of (5.15), with $\alpha = m_{12}\theta / s$ and $\beta = m_{21}\theta / s$. The distribution of sites is exchangeable (since, conditional on the coalescent tree, mutations are laid down independently at each site; this is a simple example of a marked Poisson process argument), and in particular the sites have identical distributions. They are not of course independent because of correlations induced by the common ancestry in the coalescent. However, some simple properties of the sequences are easy to calculate. In particular, the number S_n of segregating sites has mean

$$\mathbf{E}[S_n] = s\mathbf{P}(\text{site is segregating}) = s(1 - g(0) - g(n)), \qquad (5.18)$$

where $g(\cdot)$ is given by (5.16).

The equation (5.18) provides a simple heuristic method for estimating the parameters of the process. First, the equilibrium base frequencies $\pi_1 = \beta / (\alpha + \beta)$ and $\pi_2 = \alpha / (\alpha + \beta)$ are estimated from the sequence data. This done, the expected fraction of sites that are not segregating is, from (5.16) and (5.18),

$$s^{-1}\mathbf{E}(s - S_n) = \frac{\Gamma(\theta_s)}{\Gamma(\theta_s + n)}\left(\frac{\Gamma(\theta_s\pi_1 + n)}{\Gamma(\theta_s\pi_1)} + \frac{\Gamma(\theta_s\pi_2 + n)}{\Gamma(\theta_s\pi_2)}\right), \qquad (5.19)$$

where $\theta_s = \theta / s$ is the per-site substitution rate. For the pyrimidine mtDNA data, the observed fraction of nonsegregating sites is 180/201 = 0.896 and the observed fractions of C (labeled 1) and T (labeled 2) bases are $\pi_1 = 0.604$ and $\pi_2 = 0.396$, respectively. Substituting these into (5.19) and solving for θ_s give the moment estimator $\tilde{\theta}_s = 0.050$. This translates into an estimate of $\alpha = 2q_{12} = 0.050 \times 0.40 = 0.02$, and an estimate of $\beta = 2q_{21} = 0.03$. The variance of the moment estimator is hard to compute explicitly, although the top-down simulation method for the coalescent could be used to simulate the process and therefore to construct empirical estimates of the variance.

A more detailed approach to rate estimation in the finite sites model is described by Lundstrom et al. (1992a). The method is based once more on the exchangeability of the distribution of base frequencies between sites with the same mutation structure. Returning to the case in which there are K possible labelings at each site, define $V_{n,x} \equiv V_{n,(x_1,x_2,\ldots,x_K)}$ to be the fraction of sites in which x_j individuals in the sample have nucleotide j at that site, for $1 \le j \le K$. The mean of $V_{n,x}$ is given by

$$\mathbf{E}(V_{n,x}) \;=\; \mathbf{P}(\mathbf{N} = \mathrm{x}) \equiv \; q(\mathrm{x})\,, \tag{5.20}$$

the right-hand side being given by (5.14) for the independent mutation model, or by the solution of the recursion (5.17) in the general case. A least squares method obtains estimates by minimizing the squared error function

$$\sum_{\mathbf{x}} (V_{n,x} - q(\mathrm{x}))^2 .$$

This moment estimator makes fuller use of the data than the estimate based on the number of segregating sites. An alternative estimator, also described in Lundstrom et al. (1992a), is based on the assumption that the sites are evolving independently. This approximation, which is reasonable for large substitution rates (where the between-sites correlations are effectively washed out), produces a likelihood function proportional to

$$\sum_{\mathbf{x}} V_{n,\mathbf{x}} \log q(\mathbf{x})\,,$$

that can then be maximized to obtain parameter estimates.

For the mtDNA pyrimidine data, the moment method and the (independent sites) maximum likelihood method gave estimates of the C to T rate as $\alpha = 2q_{12} = 0.02$, and the T to C rate as $\beta = 2q_{21} = 0.03$. These are in close agreement with the segregating sites estimator described above. To assess the variability in the estimates of α and β, we used the top-down simulation described above, arriving at empirical bootstrap confidence intervals of $(0.01, 0.04)$ for α and $(0.02, 0.06)$ for β. These rates correspond to substitution probabilities of between 17×10^{-6} and 33×10^{-6} per site per generation for transitions from C to T, and between 25×10^{-6} and 50×10^{-6} per site per generation for transitions from T to C.

The adequacy of these estimates depends, of course, on how well the model fits the data. To assess this, we investigated how well key features of the data are reflected in simulations of the coalescent process with the given estimated rates. As might be expected, the overall base frequencies and the number of segregating sites observed in the data are accurately reflected in the simulations. One poor aspect of the fit concerned the number of distinct sequences observed in the simulations (9 to 17 per sample) compared with the 24 observed in the data. There are several reasons why such a poor fit might be observed, among them (a) site-specific variability in mutation rates, (b) admixture between genetically distinct tribes, and (c) fluctuations in population size that are not captured in the model. Further discussion of these points can be found in Lundstrom et al. (1992a) and in the final section of the present chapter.

At this point, we have come to our mathematical vignette, where population genetics theory intersects with an interesting area in combinatorics. The mathematical level of the vignette is somewhat higher than our discussion of the coalescent; readers primarily interested in aspects of the coalescent might feel justified in skipping to the final section.

MATHEMATICAL VIGNETTE: APPROXIMATING COMBINATORIAL STRUCTURES

Our mathematical vignette takes us from the world of population genetics to that of probabilistic combinatorics. We show how the Ewens sampling formula (ESF), whose origins in population genetics were described above, plays a central role in approximating the probabilistic structure of a class of combinatorial models. This brief account follows Arratia and Tavaré (1994), to which the interested reader is referred for further results. Our first task is to describe the combinatorial content of the ESF itself.

Approximations for the Ewens Sampling Formula

First, we recall Cauchy's formula for the number $N(\mathbf{a}) \equiv N(a_1, a_2, \ldots, a_n)$ of permutations of *n* objects that have a_1 cycles of length 1, a_2 cycles of length 2, . . . , a_n cycles of length *n*:

$$N(\mathbf{a}) = \mathbf{1}\left(\sum_{l=1}^{n} l a_l = n\right) n! \prod_{j=1}^{n} \left(\frac{1}{j}\right)^{a_j} \frac{1}{a_j!}, \tag{5.21}$$

$\mathbf{1}(A)$ denoting the indicator of the event *A*. If each of the *n*! permutations is assumed to be equally likely, then a random permutation has cycle index **a** with probability

$$\begin{aligned} \mathbf{P}(C_1 = a_1, C_2 = a_2, \ldots, C_n = a_n) &= \frac{N(\mathbf{a})}{n!} \\ &= \mathbf{1}\left(\sum_{l=1}^{n} l a_l = n\right) \prod_{j=1}^{n} \left(\frac{1}{j}\right)^{a_j} \frac{1}{a_j!} \end{aligned} \tag{5.22}$$

where $C_j \equiv C_j(n)$ is the number of cycles of size *j* in the permutation. Comparison with (5.4) shows that $(C_1, C_2, \ldots, C_n)$ has the ESF with parameter $\theta = 1$. To give the permutation representation of the ESF for arbitrary θ, we need only suppose that for some $\theta > 0$,

$$\mathbf{P}(\pi) = c\theta^{|\pi|}, \; \pi \in S_n , \tag{5.23}$$

where $|\pi|$ denotes the number of cycles in the permutation $\pi \in S_n$, the set of permutations of n objects. The parameter c is a normalizing constant, which may be evaluated as follows. The number of permutations in S_n with k cycles is $\left|S_n^k\right|$, the absolute value of the Stirling number of the first kind. Hence

$$1 = \sum_{\pi \in S_n} \mathbf{P}(\pi) = \sum_{k=1}^{n} \sum_{\pi:\, |\pi|=k} \mathbf{P}(\pi) = c\sum_{k=1}^{n} \left|S_n^k\right| \theta^k = c\theta_{(n)} ,$$

so that $c^{-1} = \theta_{(n)}$. It follows that under this model,

$$\begin{aligned} &\mathbf{P}(C_1 = a_1, C_2 = a_2, \ldots, C_n = a_n) \\ &\quad = \mathbf{1}\left(\sum_{l=1}^{n} l a_l = n \right) \frac{n!}{\theta_{(n)}} \prod_{j=1}^{n} \left(\frac{\theta}{j} \right)^{a_j} \frac{1}{a_j!} . \end{aligned} \tag{5.24}$$

In summary, we have shown that θ-biasing a random permutation gives the ESF.

The next ingredient in our story is the observation that the law in (5.24) can be represented as the joint law of independent Poisson random variables $Z_1, Z_2, \ldots, Z_n$, having $\mathbf{E}[Z_j] = \theta / j$, conditional on $T \equiv \sum_{j=1}^{n} jZ_j = n$:

$$\mathcal{L}(C_1, C_2, \ldots, C_n) = \mathcal{L}(Z_1, Z_2, \ldots, Z_n | T = n) , \tag{5.25}$$

where $\mathcal{L}$ denotes the law. This follows because

$$\mathbf{P}(Z_1 = a_1, Z_2 = a_2, \ldots, Z_n = a_n | T = n) = \frac{\mathbf{1}\left(\sum_{l=1}^{n} l a_l = n\right)}{\mathbf{P}(T = n)} \prod_{j=1}^{n} \frac{e^{-\theta/j} (\theta / j)^{a_j}}{a_j!},$$

which agrees with (5.24) apart from a norming constant that does not vary with $a_1, a_2, \ldots, a_n$; since both formulas are probabilities, the norming constants must be equal.

Equation (5.25) suggests that we might usefully approximate the dependent random variables $C_1, C_2, \ldots, C_n$ by the independent random variables $Z_1, Z_2, \ldots, Z_n$. This turns out to be too ambitious, but we can get away with just a little less. For any $b \in [n] \equiv \{1, 2, \ldots, n\}$, we can approximate the joint laws of $\mathbf{C}_b \equiv \mathbf{C}_b(n) \equiv (C_1, C_2, \ldots, C_n)$ by those of $\mathbf{Z}_b \equiv (Z_1, Z_2, \ldots, Z_b)$, with an error that tends to 0 as $n \to \infty$ as long as $b = o(n)$, that is, $b / n \to 0$.

As our measure of how well such an approximation might be expected to work, we use total variation distance d_{TV} as a metric on the space of (discrete) probability measures. Three equivalent definitions of the total variation distance $d_b(n)$ between (the law of) $\mathbf{C}_b$ and (the law of) $\mathbf{Z}_b$ are given below:

$$\begin{aligned} d_b(n) &\equiv d_{TV}(\mathsf{L}(\mathbf{C}_b(n)), \mathsf{L}(\mathbf{Z}_b)) \\ &= \sup_{A \subseteq N_o^b} \left| \mathbf{P}(\mathbf{C}_b(n) \in A) - \mathbf{P}(\mathbf{Z}_b \in A) \right| \\ &= \frac{1}{2} \sum_{a \in N_o^b} \left| \mathbf{P}(\mathbf{C}_b(n) = a) - \mathbf{P}(\mathbf{Z}_b = a) \right| \\ &= \inf_{couplings} \mathbf{P}(\mathbf{C}_b(n) \neq \mathbf{Z}_b). \end{aligned} \tag{5.26}$$

In (5.26), the infimum is taken over all couplings of $\mathbf{C}_b$ and $\mathbf{Z}_b$ on a common probability space, and $N_o \equiv \{0, 1, 2, \ldots\}$. Arratia et al. (1992) use a particular coupling to show that there is a universal constant $c = c(\theta)$ with $c(1) = 2$ such that

$$d_b(n) \leq c(\theta)\frac{b}{n}, \tag{5.27}$$

so that indeed $\mathbf{C}_b$ and $\mathbf{Z}_b$ can be coupled closely if (and, it turns out, only if) $b = o(n)$.

Combinatorial Assemblies

The spirit of the approximations in the preceding subsection— replacing a dependent process with an independent one—carries over to other combinatorial structures. The first of these is the class of assemblies. Assemblies are labeled structures built as follows. The set $\{1,2,\ldots,n\}$ is partitioned into a_k subsets of size *k*, for $k = 1,2,\ldots,n$, and each subset of size k is marked as one of m_k indecomposable components of size *k*. For example, in the case of permutations, $m_k = (k-1)!$, and the components of size k are the cycles on k elements. The number of structures $N(\mathbf{a})$ of weight n having a_i components of size *i*, $i = 1,2,\ldots,n$, is therefore given by

$$N(\mathbf{a}) = \mathbf{1}\left(\sum_{l=1}^{n} l a_l = n\right) n! \prod_{j=1}^{n} \left(\frac{m_j}{j!}\right)^{a_j} \frac{1}{a_j!}, \tag{5.28}$$

and the total number $p(n)$ of structures of weight n is given by

$$p(n) = \sum_{\mathbf{a}} N(\mathbf{a}). \tag{5.29}$$

A random structure of weight n is obtained by choosing one of the $p(n)$ possibilities with equal probability. If $C_j \equiv C_j(n)$ denotes the number of components of size *j*, then

$$\mathbf{P}(C_1 = a_1, C_2 = a_2, \ldots, C_n = a_n) = \frac{N(\mathbf{a})}{p(n)}. \tag{5.30}$$

In the case of permutations, this reduces to (5.22), because then $m_j / j! = 1 / j$. Note that for any $x > 0$, the probability above is proportional to

$$\mathbf{1}\left(\sum_{l=1}^{n} l a_l = n\right) \prod_{j=1}^{n} \left(\frac{m_j x^j}{j!}\right)^{a_j} \frac{1}{a_j!},$$

so that by comparison with (5.22) we see that $\mathcal{L}(C_1, C_2, \ldots, C_n) = \mathcal{L}(Z_1, Z_2, \ldots, Z_n | T = n)$, where the Z_i are independent Poisson random variables with means

$$\mathbf{E}[Z_i] = \frac{m_i x^i}{i!}, \quad i = 1, 2, \ldots, n.$$

In particular this implies that

$$\begin{aligned} d_b(n) &= d_{TV}(\mathcal{L}(\mathbf{C}_b), \mathcal{L}(\mathbf{Z}_b)) \\ &= d_{TV}(\mathcal{L}(R_b), \mathcal{L}(R_b | T = n)), \end{aligned}$$

where $R_b = \sum_{1 \le i \le b} i Z_i$. This observation reduces the calculation of a total variation distance between two processes to the calculation of a total variation distance between two random variables. We focus our attention on the class of assemblies that satisfies the logarithmic condition

$$\frac{m_i}{i!} \sim \frac{\kappa y^i}{i}, \quad i \to \infty \tag{5.31}$$

for some $\kappa, y > 0$. Among these are random permutations (for which (5.31) holds identically in i with $\kappa = y = 1$), and random mappings of $[n]$ to itself, for which $m_i = (i-1)! \sum_{j=0}^{i-1} i^j / j!$, $\kappa = 1/2, y = e$. The study of random mappings has a long and venerable history in the combinatorics literature and is reviewed in Mutafciev (1984), Kolchin (1986), and Flajolet and Odlyzko (1990), for example.

For the logarithmic class we may choose $x = y^{-1}$, and then it is known (under an additional mild rate of convergence assumption in (5.31)) that

$$d_b(n) = O\left(\frac{b}{n}\right), \tag{5.32}$$

just as for the ESF. But more detailed information is available. For example, Arratia et al. (1994b) show that for fixed b,

$$d_b(n) \sim \frac{1}{2n}|\kappa - 1|\,\mathbf{E}[|R_b - \mathbf{E}[R_b]|]. \tag{5.33}$$

The term $|\kappa - 1|$ reflects the similarity of the structure to an ESF with parameter κ, whereas the term $\mathbf{E}[|R_b - \mathbf{E}[R_b]|]$ reflects the local behavior of the structure.

The θ-biased structures, those with probability proportional to θ to the number of components, can also be studied in this way. In particular (5.30) holds, the Poisson-distributed Z_i now having mean

$$\mathbf{E}[Z_i] = \frac{\theta x^i m_i}{i!}.$$

The accuracy of the approximation of $\mathbf{C}_b$ by $\mathbf{Z}_b$ for the logarithmic class is still measured by (5.32) and (5.33), with κ replaced by $\theta\kappa$.

A rather weak consequence of the bounds typified by (5.32) and (5.33) is the fact that for each fixed b, $(C_1(n), C_2(n), \ldots, C_b(n)) \Rightarrow (Z_1, Z_2, \ldots, Z_b)$, meaning that the component counting process $\mathbf{C}$ converges in distribution (in $\mathbf{R}^\infty$) to the independent process $\mathbf{Z}$. For each n, we are comparing the combinatorial process to a *single* limiting process. This recovers the classical result of Goncharov (1944) showing that the cycle counts of a random permutation are asymptotically independent Poisson random variables with means $1/i$. The analog for random mappings is due to Kolchin (1976).

There are many uses to which such total variation estimates can be put. In essence, functionals of the dependent process that depend mainly on the small component counts (that is, on components of size $o(n)$) are well approximated by the corresponding functionals of the independent process, which are often much easier to analyze. A typical example shows that the total number of components in such a structure asymptotically has a normal distribution, with mean and variance $\theta\kappa \log n$. A corresponding functional central limit theorem follows by precisely the same methods. In addition, these estimates lead to bounds on the distances between the laws of such functionals. Some examples that illustrate the power of this approach can be found in Arratia and Tavaré (1992) and Arratia et al. (1993).

Other Combinatorial Structures

The strategy employed for assemblies also works for other combinatorial structures, including multisets and selections. We focus just on the multiset case. To build such structures, which are now unlabeled, imagine a supply of m_j types of irreducible component of weight j, and build an object of total weight n by choosing components with replacement. The number $N(\mathbf{a})$ of structures of weight n having a_j components of size $j = 1, 2, \ldots, n$ is

$$N(\mathbf{a}) = \prod_{j=1}^{n} \binom{a_j + m_j - 1}{a_j} 1\left(\sum_{l=1}^{n} l a_l = n \right), \tag{5.34}$$

and the total number of structures of weight n is $p(n) = \Sigma_{\mathbf{a}} N(\mathbf{a})$. A random multiset of size n has a_j components of size j with probability

$$\frac{1}{p(n)} \prod_{j=1}^{n} \binom{a_j + m_j - 1}{a_j} \mathbf{1}\left(\sum_{\mathbf{l=1}}^{\mathbf{n}} l a_l = n \right). \tag{5.35}$$

The ingredient common to assemblies and multisets is the fact that

$$\mathcal{L}(C_1, C_2, \ldots, C_n) = \mathcal{L}(Z_1, Z_2, \ldots, Z_n | T = n),$$

but the approximating independent random variables $\{Z_i\}$ are no longer Poisson, but rather negative binomial with parameters m_i and x^i:

$$\mathbf{P}(Z_i = k) = \binom{m_i + k - 1}{k}(1 - x^i)x^{ik}, \quad k = 0, 1, \ldots, \tag{5.36}$$

valid for $0 < x < 1$. In the θ-biased case, the Z_i are negative binomial with parameters m_i and θx^i, for any $\theta < x^{-1}$.

The most studied example in this setting concerns the factorization of a random monic polynomial over the finite field *GF*(*q*) with *q* elements. The components of size *i* are precisely the irreducible factors of degree *i*, there being

$$m_i = \frac{1}{i}\sum_{j|i} \mu(i/j) q^j$$

of them. The function $\mu(\cdot)$ is the Möbius function: $\mu(k) = -1$ or 1 according to whether *k* is the product of an odd or even number of distinct prime factors, and $\mu(k) = 0$ if *k* is divisible by the square of a prime. The logarithmic condition

$$m_i \sim \frac{\kappa y^i}{i}, \quad i \to \infty \tag{5.37}$$

is satisfied by random polynomials with $\kappa = 1$ and $y = q$. For this logarithmic class the total variation estimates (5.32) and (5.33) apply once more (with appropriate modification for the θ-biased case), and the techniques described at the end of the previous section can then be used to study the behavior of many interesting functionals. In particular, examples describing the functional central limit theorem, with error estimates, for the random polynomial case, can be found in Arratia et al. (1993).

The Large Components

Thus far we have described how we might approximate a complicated dependent process (the counts of small components) by a simpler, independent process, with an estimate of the error involved. It is natural to ask what can be said about the large component counts. To describe this, we return once more to the ESF.

Let $L_1 \equiv L_1(n) \geq L_2 \geq \cdots \geq L_K$ denote the sizes of the largest cycle, the second largest cycle, . . . , the smallest of the random number of cycles in a θ-biased random permutation. We define $L_j \equiv L_j(n) = 0, j > K$. It is known from the work of Kingman (1974, 1977) that the random vector $n^{-1}(L_1, L_2, \ldots, L_K, 0, 0, \ldots)$ converges in distribution to a random vector $(X_1, X_2, \ldots)$. The vector **X** has the Poisson-Dirichlet distribution with parameter θ, which we denote by PD(θ). There are a number of characterizations of PD(θ), among them Kingman's original definition: Let $\sigma_1 \geq \sigma_2 \geq \cdots \geq 0$ denote the points of a Poisson process on $(0,\infty)$ having mean measure with density $\theta\, e^{-x} / x, x > 0$, and set $\sigma = \sum_{i \geq 1} \sigma_i$. Then

$$\mathcal{L}(X_1, X_2 \ldots) = \mathcal{L}\left(\frac{\sigma_1}{\sigma}, \frac{\sigma_2}{\sigma}, \ldots\right).$$

We know that the large components, those that are of a size about *n*, of a θ-biased random permutation are described asymptotically by the PD(θ) law. What can be said about the large components of the other combinatorial structures we have seen? We focus once more on the logarithmic structures that satisfy either condition (5.31) or (5.37), where population genetics has a crucial role to play once more.

In approximating the behavior of counts of large components $\mathbf{C}^r \equiv (C_{r+1}, C_{r+2}, \ldots, C_n)$ we should not expect to be able to compare to an independent process because, for example, there can be at most $\lfloor n/j \rfloor$ components of size *j* or greater, and this condition forces very strong correlations on the counts of large components. However, we should be able to compare the component counting process $\mathbf{C}^r$ of the combinatorial structure to the ESF process $\hat{\mathbf{C}}^r \equiv (\hat{C}_{r+1}, \hat{C}_{r+2}, \ldots, \hat{C}_n)$, say. The approximating process is still discrete and, although not independent, it has a simpler structure than the original process. For random polynomials, it is shown in Arratia et al. (1993) that

$$d_{TV}(\mathcal{L}(\mathbf{C}^r), \mathcal{L}(\hat{\mathbf{C}}^r)) = O\left(\frac{1}{r}\right), \tag{5.38}$$

so that the counts of factors of large degree can indeed be compared successfully to the corresponding counts for the ESF. The estimate in (5.38) has as a consequence the fact that the (renormalized) factors of largest degree have asymptotically the PD(1) law, a result that also follows from work of Hansen (1994). In addition, a rate of convergence is also available. In fact, (5.38) essentially holds for any of the logarithmic class (cf. Arratia et al., 1994a).

In conclusion, we have seen that a variety of interesting functionals of the component structure of certain combinatorial processes can be approximated in total variation norm by either those functionals of an independent process or those functionals of the ESF itself. The important aspect of this is the focus on discrete approximating processes, rather than those found by renormalizing to obtain a continuous limit. In a very real sense, our knowledge of "the biology of random permutations," as described by the ESF, has provided a crucial ingredient in one area of probabilistic combinatorics.

WHERE TO NEXT?

In the preceding sections, we have illustrated how coalescent techniques can be used to model the evolution of samples of selectively neutral DNA sequence data. Simple techniques for estimating substitution rates, some based on likelihood methods and some on more ad hoc moment methods, were reviewed. We also illustrated how the probabilistic structure of the coalescent might be used to simulate observations in order to assess the variability of such estimators.

Likelihood Methods

Notwithstanding the lack of recombination and selection, inference about substitution rates poses some difficult statistical and computational problems. Most of these are due to the apparently heterogeneous nature of the substitution process in different regions of the sequence. One of the outstanding open problems in this area is the development of practical likelihood methods for sequence data. Inference techniques for sequence data from a fixed (but typically unknown) tree are reviewed in Felsenstein (1988). The added ingredient in the population genetics setting is the random nature of the coalescent itself—in principle, we have to average likelihoods on trees over the underlying coalescent sample paths. The computational problems involved in this are enormous. The likelihood can be thought of as a sum (over tree topologies) of terms, in each of which the probability of the configuration of alleles, given the branching order and coalescence times $T_n, T_{n-1}, \ldots, T_2$, is averaged over the law of $T_n, T_{n-1}, \ldots, T_2$. Monte Carlo techniques might be employed in its evaluation. One approach, using a bootstrap technique, is described by Felsenstein (1992).

An alternative approach is to compute likelihoods numerically using the recursion in equation (5.17). The probabilistic structure of the coalescent takes care of the integration, and the problem is, in principle at least, simpler. For small sample sizes and simple mutation schemes this is possible (see Lundstrom (1990), for example), but it is computationally prohibitive even for samples of the size discussed earlier. An alternative is the Markov chain Monte Carlo approach in Griffiths and Tavaré (1994a), in which equation (5.17) is used to construct an absorbing Markov process in such a way that the probability $q(\mathbf{n})$ in (5.17) is the expected value of a functional of the process up to the absorption time. That is, represent $q(\mathbf{n})$ as

$$q(\mathbf{n}) = \mathbf{E}_{\mathbf{n}}\left[\prod_{j=0}^{\tau} f(\mathbf{N}(j))\right], \qquad (5.39)$$

where $\{\mathbf{N}(j), = 0,1,\ldots\}$ is a stochastic process determined by (5.17), and τ is the time it takes this process to reach a particular set of states. Classical simulation methodology can now be used to simulate independent observations with mean $q(\mathbf{n})$. The scheme in (5.39) can be modified to estimate the entire likelihood surface from a single run, providing a computationally feasible method for approximating likelihood surfaces.

As an illustration, we return to the mitochondrial data described in the subsection on the infinitely-many-sites model above. We saw that of the 21 segregating sites in the sample, 14 were consistent with an infinitely-many-sites model. The remaining 7 sites are described in Table 5.2. These data comprise a sample of 63 individuals from a $K = 2^7 = 128$ allele model. The allele frequencies are given in Table 5.2.

Table 5.2 Incompatible Sites and Frequencies

Sequence	Site	1	9	10	13	17	18	19	Frequency
	0	T	T	C	C	T	T	C	
1		0	0	0	1	0	0	0	8
2		0	0	0	0	0	0	0	12
3		1	0	0	0	0	0	0	3
4		0	1	0	0	0	0	0	12
5		0	0	0	1	1	0	0	2
6		0	0	1	0	0	0	1	1
7		0	0	0	0	1	1	0	1
8		0	0	0	0	0	1	0	9
9		1	0	0	0	0	1	0	3
10		0	0	1	0	0	1	0	1
11		0	0	0	0	0	1	1	7
12		0	0	1	0	0	1	1	3
13		0	1	1	0	0	1	1	1

NOTE: Data are from Table 5.1. The row labeled 0 gives the nucleotide corresponding to 0 at that site. The last column gives the frequencies of the alleles in the sample.

The observed fraction of T nucleotides is $\pi_T = 207/441 = 0.469$, and so $\pi_C = 0.531$. We use these to determine the per-site mutation rate matrix Q in (5.15):

$$Q = \frac{1}{2}\begin{pmatrix} -\alpha & \alpha \\ \beta & -\beta \end{pmatrix} \equiv \frac{\theta}{2s}\left[\begin{pmatrix} \pi_{\mathrm{C}} & \pi_{\mathrm{T}} \\ \pi_{\mathrm{C}} & \pi_{\mathrm{T}} \end{pmatrix} - \begin{pmatrix} 1 & 0 \\ 0 & 1 \end{pmatrix}\right],$$

where $s = 7$. Assuming that π_{C} and π_{T} are given by their observed frequencies, there is just the single parameter θ to be estimated. Preliminary simulation results give the maximum likelihood estimate of θ at about $\hat{\theta} = 17$. This corresponds to a *per site* C $\rightarrow$ T rate of $\alpha = 1.14$, and a *per site* T $\rightarrow$ C rate of $\beta = 1.28$. These rates are about 50 times higher than those based on the analysis in the section on the *K*-allele models above using all 201 sites. Of course, this set of sites was chosen essentially because of the high mutation rates in the region and so should represent an extreme estimate of the rates in the whole molecule. Nonetheless, the results do point to the lack of homogeneity in substitution rates in this molecule. For other approaches to the modeling of hypervariable sites, see Lundstrom et al. (1992b).

Discussion

The emphasis in this chapter has been the discussion of inference techniques for the coalescent, a natural model for the analysis of samples taken from large populations.

An interesting development in the mathematical theory has been the study of measure-valued diffusions initiated by Fleming and Viot (1979). This is a generalization of the "usual" diffusions so prevalent in the classical theory of population genetics, described for example in Ewens (1979, 1990) and Tavaré (1984). A comprehensive discussion of the Fleming-Viot process appears in Ethier and Kurtz (1993), where the probabilistic structure of a broad range of examples, such as multiple loci with recombination, infinitely many alleles with selection, multigene families, and migration models, are discussed in some detail.

Perhaps the most important aspect of the theory that has seen rather little theoretical treatment thus far is the area that might loosely be called variable population size processes, and their inference. These issues are becoming more important in the analysis and interpretation of human mitochondrial sequence data. Two recent articles in this area are Slatkin and Hudson (1991) and Rogers and Harpending (1992). Lundstrom et al. (1992b) note that the effects of variable population size on gene frequency distributions can readily be confounded with the effects of hypervariable regions in the sequences. A careful assessment of the interaction of these two effects seems important, as does a detailed treatment of the effects of spatial structure and population subdivision on the analysis of sequence diversity. The Monte Carlo likelihood methods developed for sequence data in Griffiths and Tavaré (1994a) adapt readily to situations like this. See, for example, Griffiths and Tavaré (1994b.) They offer a practical approach to inference from very complicated stochastic processes. These techniques are based on genealogical arguments that provide the cornerstone of a firm quantitative basis for the analysis of DNA sequence data and our understanding of genomic diversity.

REFERENCES

General-Purpose References

Arratia, R., and S. Tavaré, 1994, "Independent process approximations for random combinatorial structures," *Adv. Math.* **104**, 90-154.

Avise, J.C., 1986, "Mitochondrial DNA and the evolutionary genetics of higher animals," *Philos. Trans. R. Soc. London, Ser. B* **312**, 325-342.

Ethier, S.N., and T.G. Kurtz, 1993, "Fleming-Viot processes in population genetics," *SIAM J. Control Optim.* **31**, 345-386.

Ewens, W.J., 1979, *Mathematical Population Genetics,* New York: Springer-Verlag.

Ewens, W.J., 1990, "Population genetics theory—the past and the future," pp. 177-227 in *Mathematical and Statistical Developments of Evolutionary Theory,* S. Lessard (ed.), Holland: Kluwer Dordrecht.

Felsenstein, J., 1982, "Numerical methods for inferring evolutionary trees," *Quarterly Review of Biology* **57**, 379-404.

Felsenstein, J., 1988, "Phylogenies from molecular sequences: inference and reliability," *Annu. Rev. Genet.* **22**, 521-565.

Hudson, R.R., 1991, "Gene genealogies and the coalescent process," pp. 1-44 in *Oxford Surveys in Evolutionary Biology* **7**, D. Futuyma and J. Antonovics (eds.).

Tavaré, S., 1984, "Line-of-descent and genealogical processes, and their applications in population genetics models," *Theor. Popul. Biol.* **26**, 119-164.

Detailed References

Anderson, S., A. Bankier, B. Barrell, M. deBruijn, A. Coulson, J. Drouin, I. Eperon, D. Nierlich, B. Roe, F. Sanger, P. Schreier, A. Smith, R. Staden, and I. Young, 1981, "Sequence and organization of the human mitochondrial genome," *Nature* **290**, 457-465.

Arratia, R., and S. Tavaré, 1992, "Limit theorems for combinatorial structures via discrete process approximations," *Random Struct. Algebra* **3**, 321-345.

Arratia, R., A.D. Barbour, and S. Tavaré, 1992, "Poisson process approximations for the Ewens sampling formula," *Ann. Appl. Probab.* **2**, 519-535.

Arratia, R., A.D. Barbour, and S. Tavaré, 1993, "On random polynomials over finite fields," *Math. Proc. Cambridge Philos. Soc.* **114**, 347-368.

Arratia, R., A.D. Barbour, and S. Tavaré, 1994a, "Logarithmic combinatorial structures," *Ann. Probab.,* in preparation.

Arratia, R., D. Stark, and S. Tavaré, 1994b, "Total variation asymptotics for Poisson process approximations of logarithmic combinatorial assemblies," *Ann. Probab.,* in press.

Cann, R., M. Stoneking, and A.C. Wilson, 1987, "Mitochondrial DNA and human evolution," *Nature* **325**, 31-36.

Di Rienzo, A., and A.C. Wilson, 1991, "Branching pattern in the evolutionary tree for human mitochondrial DNA," *Proceedings of the National Academy of Sciences USA* **88**, 1597-1601.

Estabrook, G.F., C.S. Johnson, Jr., and F.R. McMorris, 1976, "An algebraic analysis of cladistic characters," *Discrete Math.* **16**, 141-147.

Ethier, S.N., and R.C. Griffiths, 1987, "The infinitely-many-sites model as a measure valued diffusion," *Ann. Probab.* **15**, 515-545.

Ewens, W.J., 1972, "The sampling theory of selectively neutral alleles," *Theor. Popul. Biol.* **3**, 87-112.

Felsenstein, J., 1992, "Estimating effective population size from samples of sequences: A bootstrap Monte Carlo integration approach," *Genet. Res. Camb.* **60**, 209-220.

Flajolet, P., and A.M. Odlyzko, 1990, "Random mapping statistics," pp. 329-354 in *Proc. Eurocrypt '89,* J.-J. Quisquater (ed.), *Lecture Notes in Computer Science* **434**, Springer-Verlag.

Fleming, W.H., and M. Viot, 1979, "Some measure-valued Markov processes in population genetics theory," *Indiana Univ. Math. J.* **28**, 817-843.

Goncharov, V.L., 1944, "Some facts from combinatorics," *Izv. Akad. Nauk. SSSR, Ser. Mat.* **8**, 3-48 [in Russian]; "On the field of combinatory analysis," *Trans. Am. Math. Soc.* **19**, 1-46.

Griffiths, R.C., 1987, "An algorithm for constructing genealogical trees," *Statistics Research Report* **163**, Department of Mathematics, Monash University.

Griffiths, R.C., 1989, "Genealogical-tree probabilities in the infinitely-many-sites model," *J. Math. Biol.* **27**, 667-680.

Griffiths, R.C., and S. Tavaré, 1994a, "Simulating probability distributions in the coalescent process," *Theor. Popul. Biol.* **46**, 131-159.

Griffiths, R.C., and S. Tavaré, 1994b, "Sampling theory for neutral alleles in a varying environment," *Philos. Trans. R. Soc. London, Ser. B* **344**, 403-410.

Griffiths, R.C., and S. Tavaré, 1994c, "Unrooted genealogical tree probabilities in the infinitely-many-sites model," *Math. Biosci.*, in press.

Hansen, J.C., 1994, "Order statistics for decomposable combinatorial structures," *Random Struct. Alg.* **5**, 517-533.

Horai, S., and K. Hayasaka, 1990, "Intraspecific nucleotide sequence differences in the major noncoding region of human mitochondrial DNA," *Am. J. Hum. Genet.* **46**, 828-842.

Kimura, M., 1969, "The number of heterozygous nucleotide sites maintained in a finite population due to steady influx of mutations," *Genetics* **61**, 893-903.

Kimura, M., and J.F. Crow, 1964, "The number of alleles that can be maintained in a finite population," *Genetics* **49**, 725-738.

Kingman, J.F.C., 1974, "Random discrete distributions," *J. R. Stat. Soc.* **37**, 1-22.

Kingman, J.F.C., 1977, "The population structure associated with the Ewens sampling formula," *Theor. Popul. Biol.* **11**, 274-283.

Kingman, J.F.C., 1982a, "On the genealogy of large populations," *J. Appl. Probab.* **19A**, 27-43.

Kingman, J.F.C., 1982b, "The coalescent," *Stoch. Processes Appl.* **13**, 235-248.

Kolchin, V.F., 1976, "A problem of the allocation of particles in cells and random mappings," *Theory Probab. Its Applic.* (Engl. Transl.) **21**, 48-63.

Kolchin, V.F., 1986, *Random Mappings,* New York: Optimization Software, Inc.

Lundstrom, R., 1990, *Stochastic Models and Statistical Methods for DNA Sequence Data,* PhD thesis, Department of Mathematics, University of Utah.

Lundstrom, R., S. Tavaré, and R.H. Ward, 1992a, "Estimating mutation rates from molecular data using the coalescent," *Proceedings of the National Academy of Sciences USA* **89**, 5961-5965.

Lundstrom, R., S. Tavaré, and R.H. Ward, 1992b, "Modelling the evolution of the human mitochondrial genome," *Math. Biosci.* **122**, 319-336.

Maddison, D.R., 1991, "African origin of human mitochondrial DNA reexamined," *Systematic Zoology* **40**, 355-363.

McMorris, F.R., 1977, "On the compatibility of binary qualitative taxonomic characters," *Bull. Math. Biol.* **39**, 133-138.

Mutafciev, L., 1984, "On some stochastic problems of discrete mathematics," pp. 57-80 in *Mathematics and Education in Mathematics* (Sunny Beach), Sophia, Bulgaria: Bulgarian Academy of Sciences.

Nei, M., 1992, "Age of the common ancestor of human mitochondrial DNA," *Mol. Biol. Evol.* **9**, 1176-1178.

Rogers, A., and H. Harpending, 1992, "Population growth makes waves in the distribution of pairwise genetic differences," *Mol. Biol. Evol.* **9**, 552-569.

Schurr, T., S. Ballinger, Y. Gan, J. Hodge, D.A. Merriwether, D. Lawrence, W. Knowler, K. Weiss, and D. Wallace, 1990, "Amerindian mitochondrial DNAs have rare Asian mutations at high frequencies, suggesting they derived from four primary maternal lineages," *Am. J. Hum. Genet.* **47**, 613-623.

Slatkin, M., and R.R. Hudson, 1991, "Pairwise comparisons of mitochondrial DNA sequences in stable and exponentially growing populations," *Genetics* **129**, 555-562.

Templeton, A.R., 1992, "Human origins and analysis of mitochondrial DNA sequences," *Science* **255**, 737-754.

Vigilant, L., R. Pennington, H. Harpending, T. Kocher, and A.C. Wilson, 1989, "Mitochondrial DNA sequences in single hairs from a South African population," *Proceedings of the National Academy of Sciences USA* **86**, 9350-9354.

Vigilant, L., M. Stoneking, H. Harpending, K. Hawkes, and A.C. Wilson, 1991, "African populations and the evolution of human mitochondrial DNA," *Science* **253**, 1503-1507.

Ward, R.H., B.L. Frazier, K. Dew, and S. Pääbo, 1991, "Extensive mitochondrial diversity within a single Amerindian tribe," *Proceedings of the National Academy of Sciences USA* **88**, 8720-8724.

Watterson, G.A., 1975, "On the number of segregating sites in genetical models without recombination," *Theoret. Popul. Biol.* **7**, 256-276.

Wright, S., 1968, *Evolution and the Genetics of Populations,* Vol. 2, Chicago: University of Chicago Press.

Chapter 6
Winding the Double Helix: Using Geometry, Topology, and Mechanics of DNA

James H. White
University of California, Los Angeles

Crick and Watson's double helix model describes the local structure of DNA, but the global structure is more complex. The DNA double helix follows an axis that is typically curved—creating a phenomenon called supercoiling, which is crucial for a wide variety of biological processes. Understanding supercoiling requires ideas from geometry and topology. In this chapter, the author discusses three key descriptors of the geometry of supercoiled DNA molecules: linking, twisting, and writhing. These quantities are related by a fundamental theorem with important consequences for experimental biology, because it allows biologists to infer any one of the quantities from measurements of the other two.

Deoxyribonucleic acid (DNA) is usually envisioned as a pair of helices, the sugar-phosphate backbones, winding around a common linear axis. In the famous model of Crick and Watson, one turn of the double helix occurs approximately every 10.5 base pairs. However, the actual structure of DNA in a cell is typically more complex: the axis of the double helix may itself be a helix or may, in general, assume almost any configuration in space. In the late 1960s it was discovered that many DNA molecules are also closed; that is, the axis as well as the two backbone strands are closed curves. (A closed curve is a curve of finite length, the "starting point" and "endpoint" of which coincide.) In this case the DNA is called closed circular or simply closed. This chapter is concerned with the geometry, topology, and energetics of closed supercoiled DNA.

Supercoiling of closed DNA is ubiquitous in biological systems. It can arise in two ways. First, it can result when DNA winds around proteins. Second, supercoiling can also result from topological constraints known as under- or overwinding, in which case the axis of the DNA usually assumes an interwound, or plectonemic, form.

Supercoiling is important for a wide variety of biological processes. For example, supercoiling is a way of storing free energy—which can be used to assist the vital processes of replication and transcription, processes that require untwisting or separation of DNA duplex strands. Thus, supercoiling helps enzymes called helicases, polymerases, and other proteins to force apart the two strands of the DNA double helix, allowing access to the genetic information stored in the base sequence. It also promotes a variety of structural alterations that lead to DNA unwinding, such as z-DNA (left-handed double helical DNA) and cruciforms (cross-shapes). In higher organisms, supercoiling helps in cellular packaging of DNA in structures called nucleosomes, in which DNA is wound around proteins called histones. It is crucial in bringing together and aligning DNA sequences in site-specific recombination. It also changes the helical periodicity (number of base pairs per turn) of the DNA double helix; such changes can alter the binding of proteins to the DNA or the phasing of recombinant sequences.

Understanding supercoiled DNA is thus essential for the understanding of these diverse processes. Numerous biological experiments—including those based on sedimentation, gel electrophoresis, electron microscopy, X-ray diffraction, nuclease digestion, and footprinting—can give information about these matters. However, mathematical methods for describing and understanding closed circular DNA are needed to explain and classify the data obtained from these experiments.

This chapter defines and elucidates the major geometric descriptors of supercoiled DNA: linking, twisting, and writhing. It applies these concepts to classify the action of the major types of cutting enzymes, topoisomerases of Type I and Type II. It then develops the differential topological invariants necessary to describe the structural changes that occur in the DNA that is bound to proteins. Other chapters in this book explore applications of topology and geometry to DNA coiling. Chapter

7 introduces the concepts necessary to describe the mechanical equilibria of closed circular DNA and gives an analysis of transitions of superhelical transitions, dealing specifically with strand separation. Chapter 8 applies the topology of knot theory to explain the action of enzymes in carrying out the fundamental process of site-specific recombination.

DNA GEOMETRY AND TOPOLOGY: LINKING, TWISTING, AND WRITHING

To understand supercoiling in DNA, we model DNA (Bauer et al., 1980; White and Bauer, 1986) in the simplest possible way that will be useful for both "open" linear DNA and closed circular supercoiled DNA wrapped around a series of proteins. Linear DNA is best modeled by a pair of cylindrical helices, **C** and **W**, representing the backbones winding right-handedly around a finite cylinder whose central axis, **A**, is a straight line (Figure 6.1a). Such DNA has a "starting point" and an endpoint. Relaxed closed circular DNA is modeled by bending the cylinder to form a closed toroidal surface in such a way that the axis, **A**, is a closed planar curve and the ends of the curves **C** and **W** are also joined (Figure 6.1b). Finally, closed supercoiled DNA can be modeled by supercoiling the toroidal surface itself (Figure 6.1c). (Closed DNA can be used to model "open" linear DNA because the reference frame is fixed at the starting point and the endpoint of open DNA even during biological changes.)

We first wish to describe the fundamental geometric and topological quantities that can be used to characterize supercoiling, namely, the three quantities linking, writhing, and twisting (White, 1989). These are quantities that can be used to measure the interwinding of the backbone strands and the compacting of the DNA into a relatively small volume.

The **linking** number is a mathematical quantity associated with two closed oriented curves. This important property is unchanged even if the two curves are distorted, as long as there is no break in either curve. For closed DNA the linking number is that of the two curves **C** and **W**. This number can therefore be changed only by single- or double-stranded breaks in the DNA. We assume that the two strands are oriented in a parallel fashion. This assumption is not consistent with the bond polarity

FIGURE 6.1(a) The linear form of the double helical model of DNA. (b) The relaxed closed circular form of DNA. (c) The plectonemically interwound form of supercoiled closed DNA.

but greatly facilitates the mathematics necessary for the description of supercoiling. Because either backbone curve can be deformed into the axis curve **A** without passing through the other, the linking number of a closed DNA is also the linking number of either backbone curve **C** or **W** with the central axis curve **A**. Therefore, we describe the linking number of a DNA in terms of the linking number of **C** with **A**.

To define the linking number, the simplest technique is to use the so-called modified projection method. The pair of curves, **A** and **C**, when viewed from a given point distant from the two curves appear to be projected into a plane perpendicular to the line of sight except that the relative overlay of crossing segments is clearly observable. Such a view gives a modified projection of the pair of curves. In any such projection, there may be a number of apparent crossings. To each such crossing is attached a signed number ±1 according to the sign convention described in Figure 6.2. If one adds all of the signed numbers associated with this projection and divides by 2, one obtains the linking number of the curves **A** and **C**, $Lk(\mathbf{A},\mathbf{C})$, which we denote for simplicity by *Lk*. In Figure 6.3 a number of simple cases are illustrated. An important fact about the linking number of a pair of curves is that it does not depend on the projection or view of the pair; that is, the total of the signed numbers for crossings corresponding to any projection is always the same. For so-called relaxed closed DNA the average number of base pairs per turn of **C** around **W** or **C** around **A** is 10.5. Thus *Lk* can be quite large for closed DNA. For a relaxed circular DNA molecule of the monkey virus SV40, which has approximately 5,250 base pairs, *Lk* is about 500, and for bacteriophage λ of about 48,510 base pairs, *Lk* is about 4,620.

The linking number of a DNA, though a topological quantity, can be decomposed into two geometric quantities: writhe *Wr* and twist *Tw*, which can be used to describe supercoiling (White, 1969). The linking number is a measure of the crossings seen in any view. These crossings can be divided into two categories, distant crossings, which occur because the DNA axis is seen to cross itself, and local crossings, which occur because of the helical winding of the backbone curve around the axis. In the former, the backbone curve of one crossing segment is seen to cross the axis of the other segment. Distant crossings are measured by writhe, and local crossings by twist. We now give precise definitions of these two quantities.

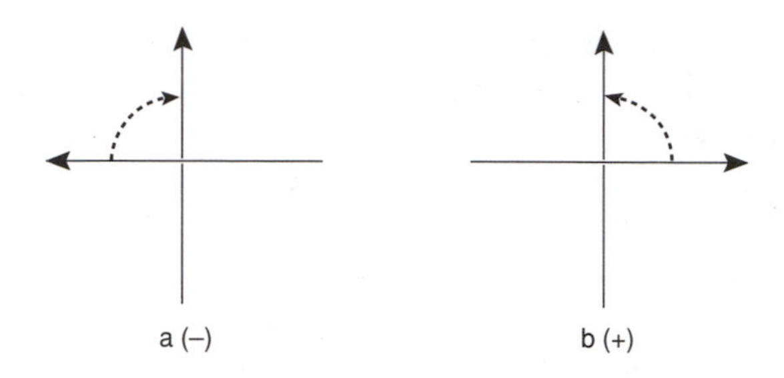

FIGURE 6.2 Sign convention for the crossing of two curves in a modified projection. The arrows indicate the orientation of the two crossing curves. To determine the sign of the crossing, the arrow on top is rotated by an angle less than 180° onto the arrow on the bottom. If the rotation required is clockwise as in (a), the crossing is given a (–) sign. If the rotation required is counterclockwise as in (b), the crossing is given a (+) sign.

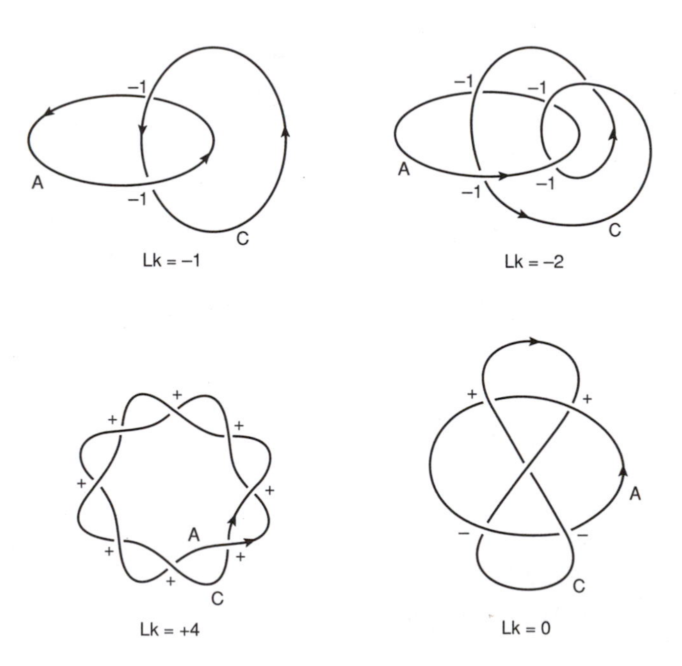

FIGURE 6.3 Examples of pairs of curves with various linking numbers, using the convention described in Figure 6.2 and the method described in the text.

Writhe can be defined in a manner analogous to the linking number. It is a property of a single curve, in this case the central axis **A**. In any modified projection of the single curve **A**, there will be a number of apparent crossings. To each such crossing is attached a sign as in the case of the linking number. If one adds all these signed numbers, one obtains the projected writhing number. Unlike the linking number, projected writhe may change depending on the view that one takes. This is demonstrated by the different views of the same curve in Figure 6.4. The writhing number is defined as the average over all possible views of the projected writhing number. If two distant segments of a DNA axis are brought very close together, then this proximity will contribute approximately ± 1 to the writhing number because in almost all views this proximity will be seen as a crossing. If the DNA axis lies in a plane and has no self-crossings, then Wr must be equal to zero, because in all views (except along the plane itself) there will be no apparent crossings. If the DNA lies in a plane except for a few places where the curve crosses itself, then the writhe is the total of the signed numbers attached to the self-crossings. Figure 6.5 gives the approximate writhe of some examples of tightly coiled DNA axes. Note that for consistently coiled curves of uniform handedness the larger the absolute value of the writhe the more compact the curve is.

An important fact about the writhe of a space curve is that if the curve is passed through itself, at the moment of self-passage the writhe changes by ± 2. This is because, at the moment of self-passage, no change takes place except at the point of passage, and the interchange of the under- and oversegments at the point of passage changes the writhe by precisely ± 2. This is illustrated in Figure 6.6. The orientation of the axis curve is not important because the writhe does not change if the orientation is reversed. This fact enables one to choose the orientation of the axis curve, **A**, to be consistent with that of the backbone strand orientation.

We next define the **twist** of the DNA. For closed DNA the twist will usually refer to the twist of one of the backbone curves, say **C** around the axis **A**, which is denoted $Tw(\mathbf{C},\mathbf{A})$ or simply Tw. To define the twist, we need the use of vectors (White and Bauer, 1986). Any local cross section of the DNA perpendicular to the DNA axis contains a unique point **a** of the axis **A** and a unique point **c** of the backbone **C**

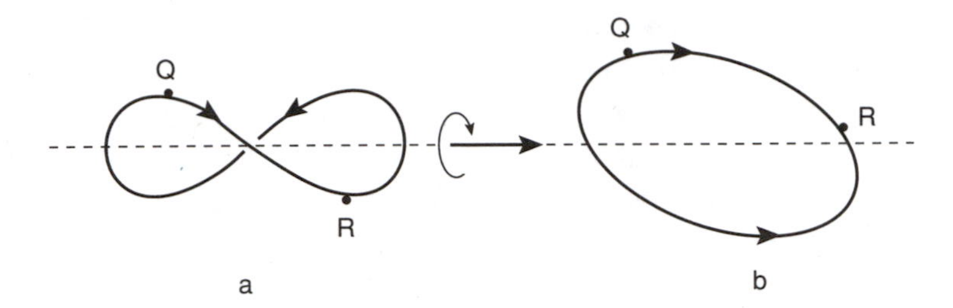

FIGURE 6.4 Illustration of the dependence of projected writhing number on projection. The axis of the same nonplanar closed DNA is shown in two different projections obtained by rotating the molecule around the dashed line. The points Q and R on the axis help illustrate the rotation. The segment QR crosses in front in (a) but is in the upper rear in (b). The projected writhing number in (a) is –1 and 0 in (b).

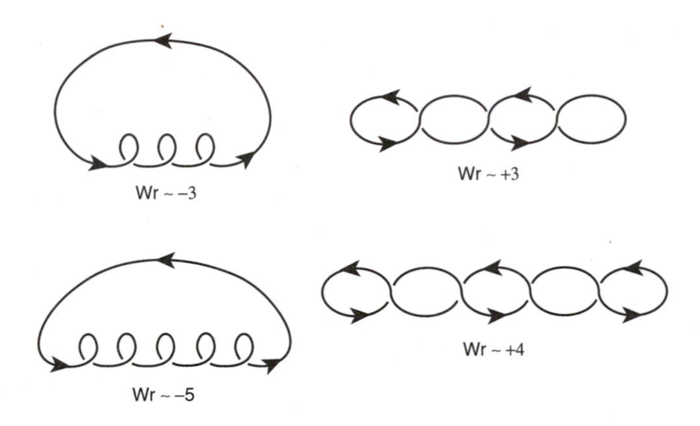

FIGURE 6.5 Examples of closed curves with different writhing numbers.

(Figure 6.7). We designate by $\mathbf{v}_{ac}$ a unit vector along the line joining the point **a** to the point **c**. Then as one proceeds along the DNA, since the backbone curve **C** turns around the axis **A,** the vector $\mathbf{v}_{ac}$ turns around the axis, or more precisely around **T**, the unit vector tangent to the curve **A**. The twist is a certain measure of this turning. As the point **a** moves along the axis **A**, the vector $\mathbf{v}_{ac}$ may change. The infinitesimal change in $\mathbf{v}_{ac}$, denoted $d\mathbf{v}_{ac}$, will have a component tangent to the axis and a component perpendicular to the axis. The twist is the measure of the total perpendicular component of the change of the vector $\mathbf{v}_{ac}$ as the point **a** traverses the entire length of the DNA. It is therefore given by the line

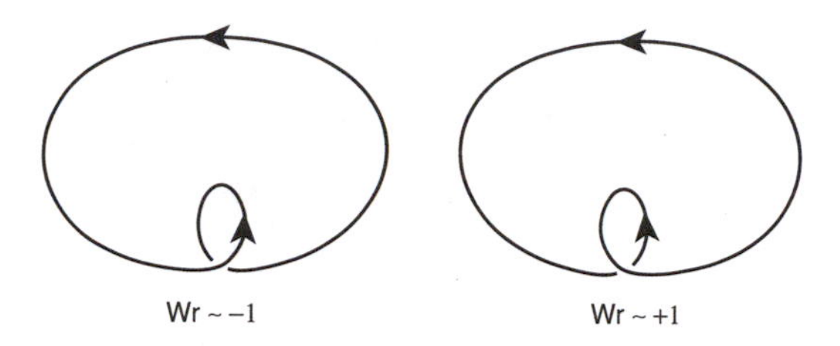

FIGURE 6.6 The writhing number of curves with one coil. The curve on the left has writhing number approximately −1 and on the right approximately +1. One curve can be obtained from the other by a self-passage at the crossing, which changes the writhing number by +2 or −2.

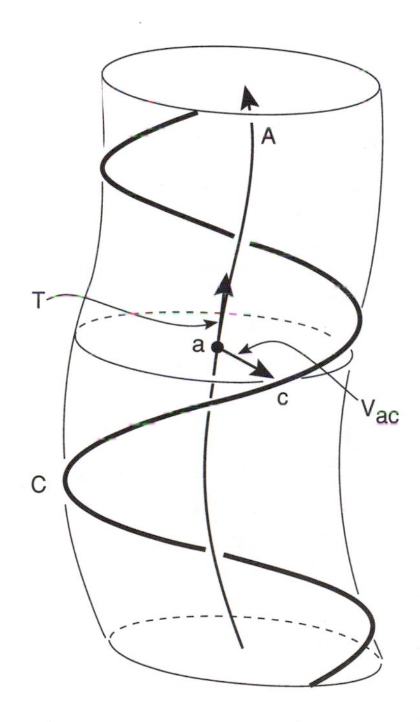

FIGURE 6.7 Cross-section of DNA. The plane perpendicular to the DNA axis **A** intersects the axis at the point **a** and intersects the backbone curve **C** at the point **c**. The unit vector along the line joining **a** to **c** is denoted $\mathbf{v}_{ac}$. Note that as the intersection plane moves along the DNA, this vector turns about the axis.

integral expression:

$$Tw = \frac{1}{2\pi} \int_{\mathbf{A}} d\mathbf{v}_{\mathbf{ac}} \cdot \mathbf{T} \times \mathbf{v}_{\mathbf{ac}} .$$

When **A** is a straight line segment or planar curve, $d\mathbf{v}_{\mathbf{ac}}$ always is perpendicular to the curve **A**, so that in these cases, *Tw* reduces to the number of times that $\mathbf{v}_{\mathbf{ac}}$ turns around the axis. Examples are shown in Figure 6.8a. Furthermore, if the DNA axis is planar and is also closed, *Tw* must necessarily be an integer, because the initial vector $\mathbf{v}_{\mathbf{ac}}$ and the final vector $\mathbf{v}_{\mathbf{ac}}$ are the same. *Tw* is not always the number of times that **C** winds around **A**. Indeed, *Tw* is usually not the number of times that **C** winds around **A** if the axis is supercoiled; Figure 6.8b gives an example in which **A** itself is a helix and **C** a superhelix winding around **A**. In this case, the twist is the number of times that **C** winds around **A** plus a term, $n \sin\gamma$, which depends on the geometry of the helix **A**. In addition, in most cases where the DNA is closed and supercoiled, the twist is not integral (White and Bauer, 1986).

The linking number, writhe, and twist of a closed DNA are related by the well-known equation (White, 1969):

$$Lk = Tw + Wr .$$

Thus for a closed strand of DNA of constant linking number, any change in *Wr* must be compensated by an equal in magnitude but opposite in sign change in *Tw*. This interchange is most easily seen by taking a rubber band or some simple elastic ribbon-like material that has two edges and while holding it fixed with one hand, twisting it with the other. After some time, much of the twisting will be seen to introduce writhing of the axis of the elastic material. The linking number of the two edges will stay the same because no breaks occur in the twisting. Because the model is held fixed by one hand, the twisting must be compensated by writhing. Though more complicated to explain, it is the same phenomenon that accounts for the supercoiling of most heavily used telephone cords. The constant twisting of the cord is eventually compensated by writhing.

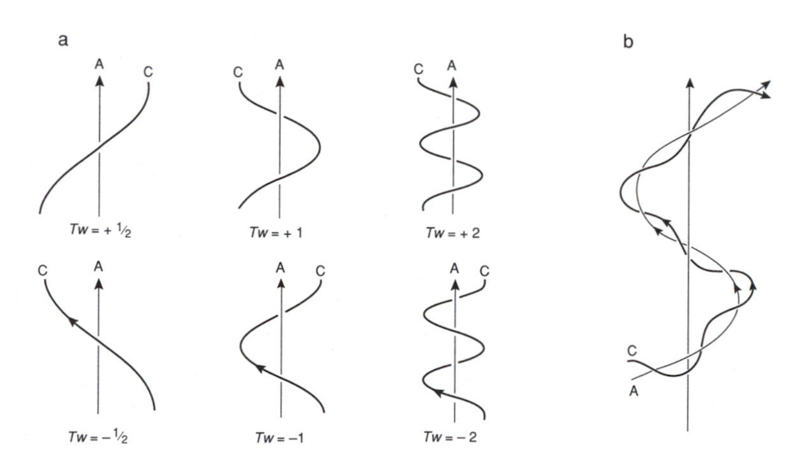

FIGURE 6.8a Examples of pairs of curves **C** and **A** with different values of twist. (a) Simple examples in which the axis **A** is a straight line and the twist is the number of times that **C** winds around **A**, being positive for right-handed twist and negative for left-handed twist. (b) An example in which the axis **A** is a helix winding around a linear axis and the curve **C** is a superhelix winding around **A**. In this case the twist of **C** around **A** is the number of times that **C** winds around **A** (in this case approximately 3.5) *plus* $n\sin\gamma$ where n is the number of times that **A** winds around the linear axis and γ is the pitch angle of the helix **A**. Here n is approximately 1.5 and γ is approximately 40°. Thus, $Tw \approx 4.46$.

APPLICATIONS TO DNA TOPOISOMERASE REACTIONS

For a relaxed closed DNA that lies in a plane and has no self-crossings, we have seen that the writhing number is equal to 0. Therefore, by the fundamental formula, $Tw = Lk$. Thus both Tw and Lk are equal to the number of times that the backbone strands wind around the axis, or more precisely the number of times the backbone curve rises above and falls below the plane in which the axis lies. For such a DNA in the B-form, there are about 10.5 base pairs per turn of the backbone. This linking number is usually denoted Lk_0. As we noted above, a relaxed DNA molecule of monkey virus SV40 with about 5,250 base pairs has $Lk_0 = Tw = 500$. However, the linking number of most closed circular DNA is not that of the relaxed state. The actual linking number

Lk in most cases is less than Lk_0. In the electron microscope, these DNA are supercoiled and appear to be contorted or coiled-up rings with many self-crossings. The quantity $Lk - Lk_0 = \Delta Lk$ is called the linking number difference and is a measure of this supercoiling. By the fundamental theorem, a change in linking must consist of a change in twist and a change in writhe: $\Delta Lk = \Delta Tw + \Delta Wr$. Because the writhe of a planar curve is 0, ΔWr becomes simply Wr. Recent work (Boles et al., 1990) shows the ratio of the change in writhe to the change in twist is approximately 2.6:1; that is, $\Delta Wr = 0.72\,\Delta Lk$ and $\Delta Tw = 0.28\,\Delta Lk$. Thus, for each change of 1 in the linking difference, there is a change of 0.72 in the writhe. Large changes in linking will therefore result in large changes in Wr. Because of this, there are negative crossovers introduced in many views of the DNA. In fact, it has been shown that the interwound coils shown in Figure 6.1c model negatively supercoiled DNA well. Such DNA are also called underwound because the twist is also reduced.

Some time after the discovery of closed supercoiled DNA (Bauer, 1978), enzymes were found that can actually change the linking number difference and in fact change a highly supercoiled DNA into a relaxed open circular DNA (Wang, 1985). These enzymes, called topoisomerases, because they change the topology of the DNA, are divided into two classes according to their operational function: Type I topoisomerases, which introduce single-stranded breaks, and Type II topoisomerases, which introduce double-stranded breaks.

An intuitive description of the action of a Type I topoisomerase can be given as follows. The first step in a Type I topoisomerase reaction is to break the backbone curve **C** of an underwound DNA at a point **c**. Then the backbone curve is allowed to rotate in a clockwise fashion to increase the twist in the direction to that preferred by B-form DNA. (This rotation is a natural process for DNA with smaller ΔLk in absolute value and is energetically more favorable.) Finally, the break is again sealed at the point **c**. The twist increases by an integral amount depending on the number of rotations. Because the axis is virtually unchanged in the immediate process, the writhe remains unchanged. Therefore the linking number increases by the number of times the **C** strand is rotated around the axis **A**. Energetically, following the completion of the sealing, each change in linking of +1 will be distributed to change the average twist by +0.28 and the average writhe by +0.72 in accordance with the results mentioned above (Boles et al., 1990). Eventually, in the presence of topoisomerases of Type I, the linking number difference of a supercoiled DNA can be reduced to 0. In reality, an entire population of supercoiled DNA with a specific linking number difference can be treated with a topoisomerase of Type I. The result will be a mixture of the same DNA with linking number differences that vary all the way from the original number to 0. By way of example, the native state of the monkey virus SV40 is not a relaxed circle, but rather a supercoiled DNA with linking number difference of −25. After treatment with Type I topoisomerase, the same population exists with linking number differences −25, −24, −23, ..., −1, 0, depending on the amount of change introduced by the enzyme. Interestingly enough, the same DNA with different linking number differences can be separated by means of gel electrophoresis (Wang and Bauer, 1979). Separation occurs due to the fact that such DNA travel through the tangled molecular matrix of a gel at different speeds because one has more *Wr* and is more compact than the other. This is one way in which the topology of a DNA can be used to characterize DNA's physical properties.

Topoisomerases of Type II also have the function of reducing the amount of linking difference. In the reaction of Type I topoisomerases described above, the twist was increased to increase the linking number. In the case of Type II topoisomerases, the increase in linking is due to an increase in writhing, which is obtained by a self-passage of the entire DNA molecule. The first step is to bring distant segments of the DNA into close proximity. Because DNA is found mostly in the interwound form, it already has distant segments reasonably close, as shown in Figure 6.1c. The node so created contributes a −1 to the writhing number at this point. Next a complete break of the DNA at the cross section of one of the neighboring segments takes place. This is essentially accomplished by breaks of both backbone chains in the cross section. The other segment is now passed through the double-stranded break. Finally, the original break is resealed. In this process of self-passage, the writhing number has been increased by +2. This is due to the fact that a negative crossing has been replaced by a positive crossing. Because the twist is virtually unaltered in the process, the linking number also increases by +2. After the process is over, the change in linking of +2 will be distributed to give a change of +1.44 in average writhe and a change of +0.56 in average twist. Continuing with the same example of SV40 DNA introduced in the paragraph above, if a topoisomerase of Type II were introduced into a population of SV40 with linking number difference −25, the result would be a collection of such DNA with linking number differences −25, −23, −21, ... , −3, −1. Such enzymes in fact were discovered because of this striking difference from topoisomerase of Type I in which DNA with all negative differences, not just the odd ones, were found.

Other functions of topoisomerases are to pass single-stranded DNA through itself or to pass nicked DNA—that is, DNA in which one of the backbone strands has been cut by an enzyme—through itself. This aspect of topoisomerases is dealt with in more detail in Chapter 8.

DNA ON PROTEIN COMPLEXES

We now turn our attention to the geometric and topological analysis of DNA whose axes are constrained to lie on surfaces (White et al., 1988). The most well-characterized example of a protein surface is the nucleosome core (Finch et al., 1977), a cylinder of height 5.04 nanometers (nm) and radius 4.3 nm. In this case the axis **A** of the DNA wraps nearly twice around the core as a left-handed helix of pitch 2.8 nm. The surface on which the DNA molecule lies is the so-called solvent-accessible surface (Richards, 1977). This is the surface generated by moving a water-sized spherical probe around the atomic surface of the protein at the van der Waals distance of all external atoms and is the continuous sheet defined by the locus of the center of the probe. (In general, the surface of a protein is defined in this manner.) It is this surface that comes into contact with the DNA backbone chain. Because the DNA is approximately 1 nm in radius, the DNA axis lies on a surface that is 1 nm outside of the solvent-accessible surface to account for the separation of the backbone from the axis. This latter surface is the one to which we shall refer in the rest of this section as the surface on which the DNA, meaning the DNA axis **A**, lies or wraps.

For DNA that lies on a surface, the geometric and topological analyses are best served by dividing the linking number not into twist and writhe, which relate only to spatial properties of the DNA, but into components that relate directly to the surface and surface-related experiments. The linking number of a closed DNA constrained to lie on the surface divides into two integral quantities, the surface linking number, which measures the wrapping of the DNA around the surface, and the winding number, which is a measure of the number of times that the backbone contacts or rises away from the surface (White et al., 1988). Experimentally, the first quantity can be measured by X-ray diffraction, and the second can be measured by digestion or footprinting methods. In particular for the nucleosome, the partial contribution to the surface linking number due to the left-handed wrapping around the cylindrical core is -1.85 (Finch et al., 1981; Richmond et al., 1984). Furthermore, the winding number has been measured to be the number of base pairs of the DNA on the nucleosome divided by approximately 10.17 (Drew and Travers, 1985).

THE SURFACE LINKING NUMBER

We now give a formal definition of the two quantities, surface linking number *SLk* and winding number, for a closed DNA on a protein surface. We assume that the surface involved has the property that at each point near the axis of the DNA, there is a well-defined surface normal vector. The unit vector along this vector will be denoted by **v**. (We assume that the surfaces are orientable. In this case, there are two possible choices for the vector field **v** depending on the side of the surface to which the vector field points.) If the DNA axis **A** is displaced a small distance $\varepsilon \neq 0$ along this vector field at each point, a new curve $\mathbf{A}_\varepsilon$ is created. ε should be chosen small enough so that during the displacement of **A** to $\mathbf{A}_\varepsilon$ no crossings of one curve with the other take place. The curve $\mathbf{A}_\varepsilon$ is also closed and can be oriented in a manner consistent with the orientation of **A**. The surface linking number is defined to be the linking number of the original curve **A** with the curve $\mathbf{A}_\varepsilon$ (White and Bauer, 1988). Simple examples of the surface linking number occur for DNA whose axes lie on planar surfaces or spheroidal surfaces. First, for a DNA whose axis lies in a plane, $SLk = 0$. This is easy to see, for in this case the vector field **v** is a constant field perpendicular to the plane. Hence the curve $\mathbf{A}_\varepsilon$ lies entirely to one side of the plane and cannot link **A**, a curve lying entirely in the plane. Second, if the DNA axis lies on a round sphere, $SLk = 0$. To see this, we can assume without loss of generality that the vector field **v** points into the sphere. In this case the displaced curve $\mathbf{A}_\varepsilon$ lies entirely inside the sphere and hence cannot link **A**. These and additional examples are illustrated in Figure 6.9.

SLk is what is technically called a differential topological invariant. As such, *SLk* has three important properties. First, if the DNA axis-surface combined structure is deformed in such a way that no discontinuities in the vector field **v** occur in the neighborhood of the DNA axis **A**, and **A** itself is not broken, then *SLk* remains invariant. For example, if the DNA lies in a plane and that plane is deformed, *SLk* remains equal to 0; if it lies on a sphere that is deformed, *SLk* remains equal to 0. Thus, if a DNA axis lies on the surface of any type of spheroid, $SLk = 0$. Examples of spheroids are shown in Figure 6.10a. An important example of *SLk* being equal to 0 is shown in Figure 6.9d, in which the DNA axis lies on the surface of a capped cylinder. A second important property of *SLk* is that it depends only on the surface near the axis. Hence, if the surface on which the axis lies is broken or torn apart at places not near the axis, *SLk* remains invariant. For example, if a DNA lies on a protein, and a portion of the protein not near the axis is broken or decomposes, *SLk* remains invariant. The third important property is that if a DNA lies on a surface and slides along the surface, then as long as the vector field **v** varies smoothly from point to point on the surface and as long as in the process of sliding the axis does not break, *SLk* remains invariant. Thus, if the capped cylinder in Figure 6.9d were allowed to expand and the axis curve required to remain the same length, it would have to unwind as it slid along the surface of the enlarged cylinder. However, *SLk* would remain equal to 0.

Another class of biologically important surfaces exists for which it is possible that a DNA can have an $SLk \neq 0$. These are so-called toroidal surfaces. They consist of the round circular torus and their deformations. Suppose an axis curve **A** traverses the entire length of a round circular torus handle once as it wraps around it a number of times, n. Suppose further that the inner radius of the torus is equal to r. For the vector field **v**, we choose the inward pointing surface normal. In this case, if one chooses $\varepsilon = r$, $\mathbf{A}_\varepsilon$ would be the central axis of the torus. Thus *SLk* is the linking number of the curve **A** with the central axis of

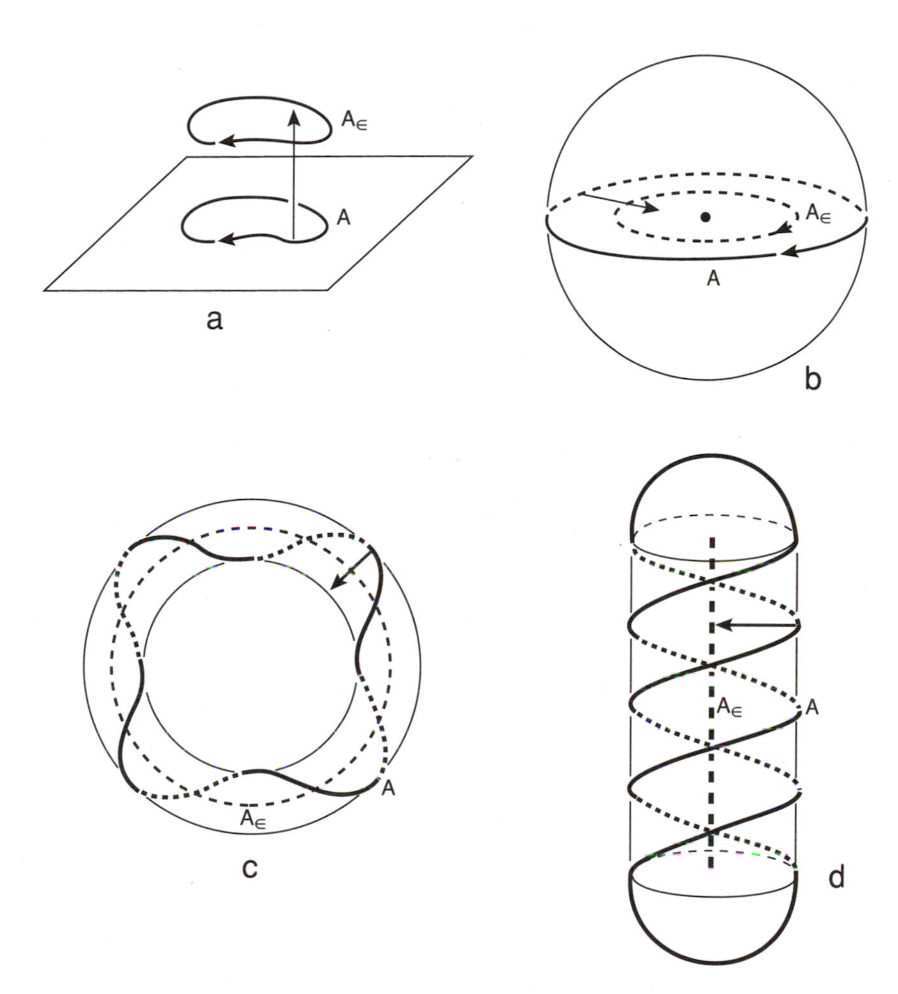

FIGURE 6.9 Examples of displacement curves and *SLk*. For any curve **A** lying on a surface, the displacement curve $\mathbf{A}_\varepsilon$ is formed by moving a small distance ε along the surface normal at each point on the curve. For planar curves as in (a), all of the normal vectors can be chosen to point upward, and then $\mathbf{A}_\varepsilon$ is above **A**. The curves are unlinked, and hence $SLk = 0$. For curves on a spherical surface as in (b), the surface vectors can be chosen to point inward, and hence $\mathbf{A}_\varepsilon$ is entirely inside and therefore does not link **A**. *SLk* is again equal to 0. In (c) and (d) the surface normal vectors have been chosen to point inward, and ε has been set equal to the inner radii of the surfaces on which the DNA is wound. In (c), $\mathbf{A}_\varepsilon$ becomes the central axis of the torus, and $SLk = +4$. In (d) the DNA is wrapped plectonemically around a capped cylinder. The displacement curve $\mathbf{A}_\varepsilon$ lies entirely inside, and thus $SLk = 0$.

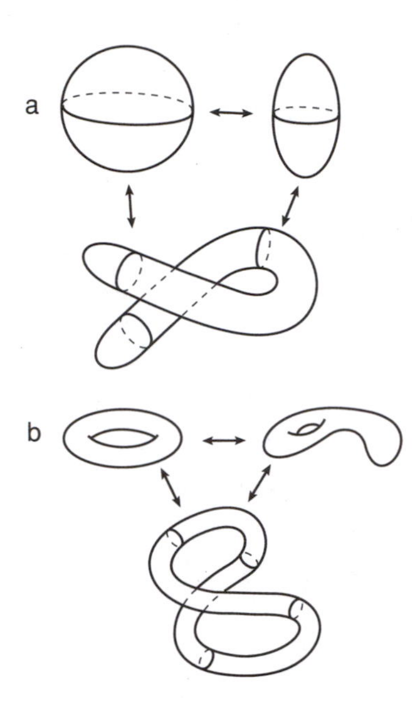

FIGURE 6.10 (a) Deformations of the round sphere into spheroids. (b) Deformation of the round circular torus into toroids.

the torus. This implies that if the wrapping is right-handed, $SLk = +n$, and if the wrapping is left-handed, $SLk = -n$. By the invariant properties mentioned above, *SLk* remains invariant even if the round torus is deformed. Examples are shown in Figure 6.10b.

The concept of *SLk* can also be applied to DNA that are not attached to real protein surfaces but are free in space. For example, the most common kind of free DNA, that is, DNA free of any protein attachment, is plectonemically wound DNA. Here the DNA can be considered to lie on the surface of a spheroid such as the one shown in Figure 6.9b (or a deformation of it), the exact shape of which is determined by the energy-minimum DNA conformation. Then the surface may be allowed to vanish and reappear without changing the shape of the DNA superhelix. The DNA is said to be wrapped on a *virtual* surface (White et al., 1988). Thus, the *SLk* of the DNA in Figure 6.9d is equal to 0 regardless of whether the surface is virtual or is that of a real protein. More generally, these concepts can be applied to DNA wrapped on a series of proteins with virtual surfaces joining them. An example of this is presented below in our discussion of the minichromosome.

THE WINDING NUMBER AND HELICAL REPEAT

We next give a formal definition of the winding number of a DNA wrapping on a surface. Because the vector **v** is perpendicular to the surface, it is also perpendicular to the DNA axis **A** and thus lies in the perpendicular planar cross section at each point of the DNA. Therefore, at each point this vector **v** and the strand-axis vector $\mathbf{v}_{ac}$ defined above lie in this same planar cross section. In this plane, the vector $\mathbf{v}_{ac}$ makes an angle ϕ with the vector **v** (Figure 6.11). As one proceeds along the DNA segment, $\mathbf{v}_{ac}$ spins around **v**, as the backbone curve **C** alternately rises away from and falls near to the surface, while the angle ϕ turns through 2π radians (that is, 360°). The total change in the angle ϕ, divided by the normalizing factor 2π (or 360°), during this passage is called the winding number of the DNA and is denoted Φ (White et al., 1988). This number may also be thought of as the number of times that $\mathbf{v}_{ac}$ rotates past **v** as the DNA is traversed. A related quantity called the helical repeat, denoted h, is the number of base pairs necessary for one complete 360° revolution. For closed DNA, because the beginning and ending point are the same for one complete passage of the DNA, the vectors **v** and $\mathbf{v}_{ac}$ are the same at the beginning and at the end. In this case, therefore, Φ must necessarily be an integer.

There are equivalent formulations for the winding number of a closed DNA wrapping on a protein surface. During each 360° rotation of the vector $\mathbf{v}_{ac}$ in the perpendicular plane, ϕ assumes the values of 0 (or 0°) and π (or 180°) exactly once. When $\phi = 0$, $\mathbf{v}_{ac} = \mathbf{v}$, and when $\phi = \pi$, $\mathbf{v}_{ac} = -\mathbf{v}$. In the latter case, the backbone strand is at maximal distance from the protein surface, and in the former it comes into contact with the protein. Thus the winding number of a closed DNA on a protein surface is the number of times one of the backbone strands contacts the protein surface, or it is the number of times the strand is at maximal distance from it. In this case, it also follows from its definition that the helical repeat is the number of base pairs between successive contact points of one of the backbones or the number of base pairs between successive

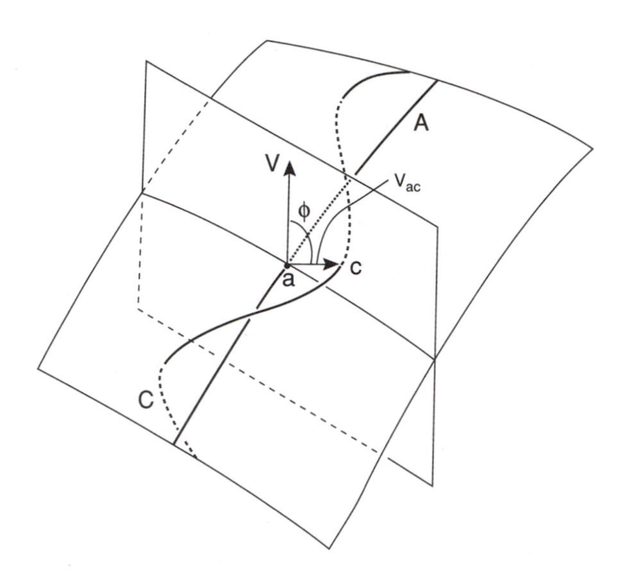

FIGURE 6.11 Definition of the surface vectors necessary to define the winding number. The duplex DNA axis **A** lies on the surface. The backbone curve **C** will pass above and below the surface as it winds around **A**. To describe surface winding, two vectors need to be defined originating at a point **a** on **A**, namely, the unit surface normal vector **v** and the strand-axis vector $\mathbf{v}_{ac}$ along the line connecting **a** to the corresponding point **c** on the backbone **C**. ϕ is defined to be the angle between these two vectors. The winding number Φ measures the number of times that ϕ turns through 360°, or how many times $\mathbf{v}_{ac}$ rotates past **v**.

points of maximal distance from the protein surface. This latter number can be measured directly by digestion or footprinting methods, which involve probes that search for points of the backbones to cut, the easiest being those points at maximal distance from the surface.

The winding number Φ is also a differential topological invariant and therefore has the same three properties mentioned above for *SLk*. In particular, it remains invariant if the DNA surface structure is deformed without any breaks in the DNA or any introduction of discontinuities in the vector field **v**. Under the same conditions, it also remains unchanged if the DNA is allowed to slide along the surface.

RELATIONSHIP BETWEEN LINKING, SURFACE LINKING, AND WINDING

It is remarkable that the three quantities *Lk*, *SLk*, and Φ, although very different in definition, are related by a theorem from differential topology. In fact, for a closed DNA on a surface, the linking number is the sum of the surface linking number and the winding number (White et al., 1988); that is,

$$Lk = SLk + \Phi .$$

Before we outline the proof of this result, we first give some simple examples. We then give the proof and conclude with the example of the minichromosome.

For DNA that lies in a plane or on a spheroid, $SLk = 0$. Therefore, $Lk = \Phi$, and if there are N base pairs in the DNA, the helical repeat is given by $h = N / Lk$. These two cases include relaxed circular DNA, for which $Lk = Lk_0$, and plectonemically interwound DNA, the most common form of supercoiled DNA. For DNA that traverses the handle of a round circular torus while wrapping n times around the handle, $Lk = n + \Phi$ if the wrapping is right-handed, and $Lk = -n + \Phi$ if the wrapping is left-handed. In both cases, *Lk* is unchanged if the torus is smoothly deformed.

We now outline the proof of the main result. To do this, we first define the surface twist, *STw*, of the vector field **v** along the axis curve **A** (White and Bauer, 1988; White et al., 1988). This is basically defined the same as the twist of the DNA except that the vector field **v** is used and not the vector field $\mathbf{v}_{ac}$. Hence, *STw* is given by the equation

$$STw = \frac{1}{2\pi} \int_{\mathbf{A}} d\mathbf{v} \cdot \mathbf{T}_a \times \mathbf{v} .$$

Thus, *STw* measures the perpendicular component of the change of the vector **v** as one proceeds along the axis **A**, and thus is a measure of the spinning of the vector field **v** around the curve **A**. It can also be considered to be the twist of $\mathbf{A}_\varepsilon$ around **A**. We recall that *Tw* measures the spinning of the vector field $\mathbf{v}_{ac}$ around the curve **A**. Thus, the difference $Tw - STw$ measures the spinning of $\mathbf{v}_{ac}$ around **v**. But this is exactly the winding number Φ. Hence $Tw - STw = \Phi$.

We recall the fundamental formula

$$Wr(\mathbf{A}) + Tw(\mathbf{C},\mathbf{A}) = Lk(\mathbf{C},\mathbf{A}) \text{ or}$$

$$Wr + Tw = Lk\,.$$

A similar formula relating *STw*, *Wr*, and *SLk* holds:

$$Wr(\mathbf{A}) + STw = Lk(\mathbf{A},\mathbf{A}_\varepsilon) \text{ or}$$

$$Wr + STw = SLk$$

because *STw* is the twist of $\mathbf{A}_\varepsilon$ around **A** and *SLk* is the linking number of $\mathbf{A}_\varepsilon$ and **A**. Combining the two formulas and using the result that $Tw - STw = \Phi$, we obtain

$$Lk = SLk + \Phi\,.$$

The biological importance of this relationship is that all three of these quantities are experimentally measurable. Thus, having determined any two of them, one can calculate the other and then compare with the experimental value. In the next section, we show by a classical example from molecular biology, the minichromosome, the power of this theorem.

APPLICATION TO THE STUDY OF THE MINICHROMOSOME

A minichromosome is a structure that consists of a closed DNA bound to a series of core nucleosomes. Such a structure allows the compaction of a very long DNA into a small volume, in the same way that a long piece of thread is compacted by wrapping it on a spool. Understanding such structures is essential to a knowledge of how DNA is packaged in the cell. In this section, we study the geometry and topology of DNA in such a structure. Each nucleosome may best be described as a cylinder, the histone octamer, around which the DNA wraps approximately 1.8 times in a left-handed manner. The DNA segments between successive nucleosomes are called linker regions. Thus, the DNA divides between linker DNA and core-associated DNA. An example of such a structure is shown in Figure 6.12. Such a compound structure consists of a toroidal surface, part of which is the real surface of the nucleosome cores and part of which is virtual linker surfaces joining successive cylinders. These virtual pieces are deformed cylindrical sections, all of the same radius, on which the linker DNA are constrained to lie. The specification of each of these surfaces is arbitrary as long as it takes into account the coiling of the linker. The linker DNA can thus be thought of as a generating curve for the cylindrical section. An important condition to be imposed is that the linker DNA does not wind around the piece on which it lies. This condition will ensure that all contributions to *SLk* due to winding around the torus handle will come only from the intranucleosome winding. Any additional contribution to *SLk* must therefore come from the coiling of the linker DNA.

To simplify our example, we will assume that the minichromosome is relaxed. This means that the linker regions are planar and that all contributions to *SLk* come from the winding of the DNA around the histone octamers. Such a relaxed state can be achieved by the introduction into the minichromosome of topoisomerases, which relax the linker DNA but leave unaffected the DNA on the nucleosome cores. In this case, *SLk* can be directly measured by X-ray diffraction and found to be $-1.8\,m$, where m is the number of nucleosomes. An example with 5 nucleosomes is shown in Figure 6.13, for which $SLk = -9$. For SV40 DNA, there are about 25 nucleosomes (Sogo et al., 1986). Therefore, $SLk = -45$.

The linking number of the DNA on the relaxed SV40 minichromosome is measured in an indirect way. First, the DNA is stripped of the nucleosome particles, becoming in the process a plectonemically interwound free DNA. By means of gel electrophoresis, its linking number can be experimentally measured. In actuality, what is measured is the difference of its linking number and the linking number of the same DNA totally relaxed, ΔLk, as defined above. ΔLk is found to be about -1 per nucleosome core; that is, $\Delta Lk = -25$ (Shure and

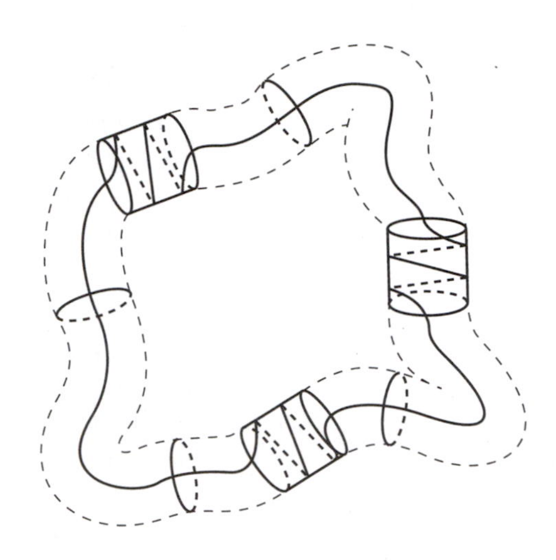

FIGURE 6.12 Cartoon of a minichromosome. Three cylinders representing histone octamers are wound by DNA so as to form three nucleosomes. The nucleosomes are connected by linker DNA segments. Successive real nucleosomes are connected by virtual deformed cylindrical pieces, the deformations of which are determined by the coiling of the linker. Reprinted, by permission, from White et al. (1988). Copyright © 1988 by American Association for the Advancement of Science.

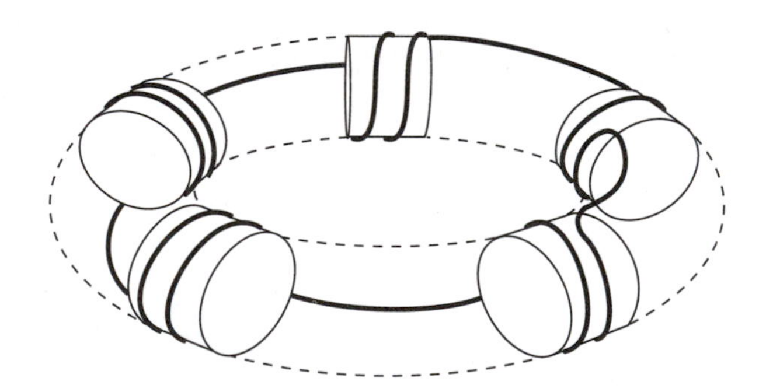

FIGURE 6.13 Diagram of a relaxed minichromosome with five cylindrical nucleosomes. The DNA wraps left-handedly 1.8 times around each nucleosome. The contribution to *SLk* is −1.8 for each nucleosome and 0 for each linker region. For the entire structure, *SLk* = −9. Reprinted, by permission, from White et al. (1989). Copyright © 1989 by Academic Press Limited.

Vinograd, 1976). As we stated above, relaxed SV40 has a linking number, Lk_0, of approximately 500. Therefore minichromosomal SV40 has $Lk = 475$.

We can now answer an important question: Is the number of base pairs per turn, the helical repeat, unchanged from the 10.5 of relaxed DNA, when DNA is wrapped on the nucleosome? The answer must be negative because of the relationship $Lk = SLk + \Phi$. Thus, we can theoretically determine that because $Lk = 475$ and $SLk = -45$, Φ must be 520. However, we have seen above that Φ for relaxed SV40 is equal to $Lk_0 = 500$. Because $\Phi = 520$, the average helical repeat for minichromosomal SV40 equals $5{,}250/520 = 10.10$. In this analysis, we have made a great many simplifications, but it is noteworthy that this number is in remarkably good agreement with the number 10.17 that is obtained by nuclease digestion experiments. The answer to the question is thus negative.

To summarize, we have found a fundamental relationship $Lk = SLk + \Phi$ for three quantities that are directly accessible to experiment, Lk by electrophoresis, SLk by X-ray diffraction, and Φ by digestion. If two of the three are known, one can use the relationship to predict and therefore verify the experimental evidence for finding the third. This gives a powerful use of differential topology in the field of molecular biology.

REFERENCES

Bauer, W.R., 1978, "Structure and reactions of closed duplex DNA," *Annu. Rev. Biophys. Bioeng.* **7**, 287-313.

Bauer, W.R., F.H.C. Crick, and J.H. White, 1980, "Supercoiled DNA," *Scientific American* **243**, 118-133.

Boles, T.C., J.H. White, and N.R. Cozzarelli, 1990, "Structure of plectonemically supercoiled DNA," *Journal of Molecular Biology* **213**, 931-951.

Drew, H.R., and A.A. Travers, 1985, "DNA bending and its relation to nucleosome positioning," *Journal of Molecular Biology* **186**, 773-790.

Finch, J.T., L.C. Lutter, D. Rhodes, R.S. Brown, B. Rushton, M. Levitt, and A. Klug, 1977, "Structure of nucleosome core particles of chromatin," *Nature* **269**, 29-36.

Finch, J.T., R.S. Brown, D. Rhodes, T. Richmond, B. Rushton, L.C. Lutter, and A. Klug, 1981, "X-ray diffraction study of a new crystal form of the nucleosome core showing higher resolution," *Journal of Molecular Biology* **145**, 757-770.

Richards, F.M., 1977, "Areas, volumes, packing and protein structure," *Annu. Rev. Biophys. Bioeng.* **6**, 151-176.

Richmond, T.J., J.T. Finch, B. Rushton, D. Rhodes, and A. Klug, 1984, "Structure of the nucleosome core particle at 7Å resolution," *Nature* **311**, 532-537.

Shure, M., and J. Vinograd, 1976, "The number of superhelical turns in native virion SV40 DNA and minicol DNA determined by the band counting method," *Cell* **8**, 215-226.

Sogo, J.M., H. Stahl, T. Koller, and R. Knippers, 1986, "Structure of replicating simian virus 40 minichromosomes. The replication fork, core histone segregation and terminal structures," *Journal of Molecular Biology* **189**, 189-204.

Wang, J.C., 1985, "DNA topoisomerases," *Annu. Rev. Biochem.* **54**, 665-697.

Wang, J.C., and W.R. Bauer, 1979, "The electrophoretic mobility of individual DNA topoisomers is unaffected by denaturation and renaturation," *Journal of Molecular Biology* **129**, 458-461.

White, J.H., 1969, "Self-linking and the Gauss integral in higher dimensions," *Am. J. Math.* **91**, 693-728.

White, J.H., 1989, "An introduction to the geometry and topology of DNA structure," pp. 225-253 in *Mathematical Methods for DNA Sequences,* M.S. Waterman (ed.), Boca Raton, Fla: CRC Press.

White, J.H., and W.R. Bauer, 1986, "Calculation of the twist and the writhe for representative models of DNA," *Journal of Molecular Biology* **189**, 329-341.

White, J.H., and W.R. Bauer, 1988, "Applications of the twist difference to DNA structural analysis," *Proceedings of the National Academy of Sciences USA* **85**, 772-776.

White, J.H., N.R. Cozzarelli, and W.R. Bauer, 1988, "Helical repeat and linking number of surface-wrapped DNA," *Science* **241**, 323-327.

White, J.H., R. Gallo, W.R. Bauer, 1989, "Effect of nucleosome distortion on the linking deficiency in relaxed minichromosomes," *Journal of Molecular Biology* **207**, 193-199.

Chapter 7
Unwinding the Double Helix: Using Differential Mechanics to Probe Conformational Changes in DNA

Craig J. Benham
Mount Sinai School of Medicine

The two strands of DNA are usually bound together in a double helix. However, many key biological processes—including DNA replication and gene expression—require unwinding of the double helix. Such unwinding requires the input of energy, a large part of which is stored in the form of supercoiling of a chromosome or chromosomal region. Given a supercoiled DNA molecule, where along its sequence will unwinding occur? In this chapter, the author shows how basic principles of statistical mechanics—together with some delicate numerical estimates—can be applied to predict the sites of supercoil-induced unwinding. The mathematical predictions are abundantly confirmed by experimental data and, when applied to new situations, they suggest novel insights about gene regulation.

Deoxyribonucleic acid (DNA) usually occurs in the familiar Watson-Crick B-form double helix, in which the two strands of the DNA duplex are held together by hydrogen bonds between their complementary bases. Many important biological processes, however, involve separating the strands of the DNA duplex in order to gain access to the information encoded in the sequence of bases within individual strands. In transcription, the first step in gene expression, the DNA base pairing within the gene must be temporarily disrupted to allow an RNA molecule with a sequence complementary to one of the strands of the gene to be constructed. In DNA replication, the two original strands of a parent DNA molecule replicate to form two complete molecules, with each strand serving as a template for the synthesis of its complement. To accomplish this, the strands of the parent molecule must separate to provide access to these templates.

The regulation of important physiological processes is extremely precise and complex. In addition to many other layers of control, the strand separations required for specific functions must be carefully regulated to occur at the precise positions needed for each activity, and only at times when that activity is to be initiated. Because DNA prefers to remain in the B-form under normal conditions, strand separation requires the expenditure of (free) energy. The energy required for strand separation depends upon the sequence of base pairs being separated. Because A·T base pairs are held by only two hydrogen bonds whereas G·C pairs are held by three, it is energetically less costly to separate the former pairs than the latter. For this reason, strand separations tend to be concentrated in A+T-rich regions of the DNA. As we will see in this chapter, this provides the sequence dependence necessary to control the sites of separation.

Controlling the occurrence of separations can be accomplished by modulating the amount of energy stored in the DNA molecule itself. This is done by changing the topological constraints on the molecule. DNA in living organisms is topologically constrained into domains within which the linking number is fixed. Enzymes can change this linking number, placing the DNA in a higher energy state in which pure B-form DNA is less favored and partial strand separation is thermodynamically more achievable. (The topology and geometry of superhelicity, which is the jargon name for this process, have been described by White in Chapter 6.)

In order to illuminate the role of strand separation in DNA functions, one needs accurate theoretical methods for predicting how a particular DNA sequence will behave as its linking number is varied. This chapter describes methods that have been developed to make such predictions. The results of sample calculations are shown, and the insights that they provide regarding specific DNA activities are sketched. The global and topological nature of the constraints imposed on DNA causes behavior that exhibits many unusual and surprising features.

DNA SUPERHELICITY—MATHEMATICS AND BIOLOGY

DNA in living cells is held in topological domains whose linking numbers can be individually regulated. In practice there are two types of domains. Small DNA molecules can occur as closed circles, whereas larger DNA molecules are formed into a series of loops by periodic attachments to a protein scaffold in a way that precludes local rotations at the attachment site. This arrangement constrains the portion of DNA between adjacent attachment sites to be a topological domain analogous to a closed circle.

For simplicity we consider a closed circular duplex DNA molecule as the paradigm of the topological domain. (Closed circles are also the molecules of choice for experiments in this field.) The two strands that make up the DNA duplex each have a chemical orientation induced by the directionality of the bonds that join neighbor bases. This is called the 5'-3' orientation because each phosphate group in a strand joins the 5' carbon of one sugar to the 3' carbon of the next. This orientation must be the same for every phosphate group within a strand, which imparts a directionality to the strand as a whole. The two strands of the B-form duplex are oriented so their 5'-3' directions are antiparallel. In consequence, a duplex DNA molecule can be closed into a circle only by joining together the ends of each individual strand. Circularization by joining the ends of one strand to those of the other to form a Möbius strip is forbidden because the bonds required would violate the conservation of 5'-3' directionality. Hence a closed circular DNA molecule is composed of two interlinked, circular (antiparallel) strands.

Circularization fixes the linking number of the resulting molecule; the linking number is the number of times that either strand links through the closed circle formed by the other strand. (Topological domains formed by periodic attachments have a functionally equivalent constraint.) The fixing of the linking number Lk within a topological domain provides a global constraint that topologically couples its secondary and tertiary structures according to White's (1988) formula

$$Lk = Tw + Wr . \tag{7.1}$$

Although *Lk* is fixed in a topological domain, both *Tw* and *Wr* may still vary, provided they do so in a complementary manner.

Cutting one DNA strand in a domain releases the topological constraint of constant *Lk*, allowing it to find its most relaxed state. The two resulting ends may rotate freely, relaxing any torsional deformation imposed on the molecule. Writhing deformations can be converted to twist and then removed by this rotational relaxation. The sum of the twist and writhe in this relaxed state determines a relaxed linking number Lk_0. Note that, while the linking number *Lk* of a circular DNA molecule must be an integer, the relaxed linking number Lk_0 need not be integral.

Stresses are imposed on a topological domain whenever its linking number *Lk* differs from the relaxed value. The resulting linking difference $\alpha = Lk - Lk_0$ must be accommodated by twisting and/or writhing deformations:

$$\alpha = \Delta Tw + \Delta Wr . \tag{7.2}$$

Topological domains in living systems are commonly found in a negatively superhelical state, in which the imposed linking number is smaller than its relaxed value, so $\alpha < 0$. Negative superhelicity provides a mechanism for driving strand separation. Because the separated strands are less twisted than the B-form, they localize some of the linking deficiency as a decrease of twist at the transition site, thereby allowing the rest of the domain to relax a corresponding amount. Since strand separations require energy, they are disfavored in unconstrained or relaxed molecules. However, in a negatively superhelical domain, local strand separations are energetically favored to occur at equilibrium whenever the topological strain energy that is thus relieved exceeds the energetic cost of locally disrupting the base pairing between strands.

The linking differences imposed on topological domains in vivo are carefully regulated. Virtually all organisms produce enzymes that alter *Lk* through the introduction of transient strand breaks (Gellert, 1981). The action of these molecules maintains topological domains in negatively superhelical, underlinked states (i.e., $\alpha < 0$). On average, bacteria and other primitive organisms maintain approximately half their domains in a superhelical state. Moreover, the amount of superhelicity imposed on DNA in vivo is known to vary with the cell division cycle in a carefully regulated manner (Dorman et al., 1988). The extent of superhelicity also varies in response to environmental changes (Bhriain et al., 1989; Malkhosyan et al., 1991). In multicelled organisms, superhelicity occurs primarily within domains containing actively expressing genes. The DNA within malignant cancer cells is maintained at more extreme negative linking differences than that characterizing the corresponding DNA in normal cells (Hartwig et al., 1981).

Many important regulatory events are sensitive to the degree of superhelical stress imposed on the DNA. These include the initiation of gene expression (Smith, 1981; Pruss and Drlica, 1989; Weintraub et al., 1986) and of DNA replication (Kowalski and Eddy, 1989; Mattern and Painter, 1979). Substantial evidence suggests that superhelically driven strand separations may be involved in these processes. One well-characterized case occurs at the origin of DNA replication of the bacterium *E. coli* (Kowalski and Eddy, 1989). The DNA sequence at this origin site contains a triple repeat of an A+T-rich run of 13 base pairs that is required for the initiation of DNA replication. Deletion and substitution experiments have shown that the key functional attribute of this sequence is its susceptibility to superhelical strand separation. DNA sequence changes at this site that retain this attribute preserve its ability to initiate replication in vivo; DNA sequence changes that degrade this susceptibility destroy in vivo origin function. No other sequence specificity is observed. Such sequences are called duplex unwinding elements (DUEs) and are present at origins of DNA replication in many organisms (Umek et al., 1989).

Superhelicity also is known to modulate the expression of some genes. In bacteria, superhelicity regulates the expression of the so-called SOS system, a suite of genes that are activated in response to environmental stresses or DNA damage. The bacterial response to deleterious environmental changes is to increase the superhelicity of its DNA, which activates expression of the SOS genes (Bhriain et al., 1989; Malkhosyan et al., 1991). Experimental (Kowalski et al., 1988) and theoretical (Benham, 1990) results indicate that the susceptibility of some DNA molecules to superhelical strand separation is confined to sites that bracket specific genes. This suggests that there may be at least two classes of genes, distinguishable by their sensitivities to superhelical separation, whose mechanisms of operation may be different.

Strand separation in living organisms frequently arises through interactive processes, in which local superhelical destabilization of the B-form acts in concert with other factors. Biological systems may exploit marginal decreases in the stability of the B-form that occur at discrete sites in superhelical molecules. For example, consider an enzyme that functions by recognizing a particular sequence and inducing separation there. It might be energetically able to induce the transition only if the B-form already is marginally destabilized at that site. This suggests that superhelical helix destabilization also can regulate biological processes through mechanisms that need not involve preexisting separations. For this reason it is important also to develop methods to predict sites where superhelicity marginally destabilizes the duplex.

STATEMENT OF THE PROBLEM

This chapter develops methods to predict the strand separation and helix destabilization experienced by a specified DNA sequence when superhelically stressed. We will focus specifically on predictions regarding several plasmids (that is, circular DNA molecules) that have been engineered to include the *E. coli* replication origin or variants thereof. This is done because experimental information is available regarding superhelical strand separation in these molecules.

In principle, the analysis of conformational equilibria is quite direct. Because every base pair can separate, there are many possible states of strand separation available to a topologically constrained DNA molecule. By basic statistical mechanics, a population of identical molecules at equilibrium will be distributed among its accessible states according to Boltzmann's law. If these states are indexed by *i*, and if the free energy of state *i* is G_i, then the equilibrium probability p_i of a molecule being in state *i*, which is the fractional occupancy of that state in a population at equilibrium, equals

$$p_i = \frac{\exp(-G_i / RT)}{Z}. \tag{7.3}$$

Here Z is the so-called partition function, given by

$$Z = \sum_i \exp(-G_i / RT), \tag{7.4}$$

where R is the gas constant and T is the absolute temperature. Thus the fractional occupancies of individual states at equilibrium decrease exponentially as their free energies increase. If a parameter ζ has value ζ_i in state i, then its population average, that is, its expected value $\bar{\zeta}$ at equilibrium, is

$$\bar{\zeta} = \sum_i \zeta_i p_i . \tag{7.5}$$

This expression can be used to evaluate any equilibrium property of interest, once the governing partition function is known.

The application of this approach to the rigorous analysis of conformational equilibria of superhelical DNA molecules is complicated by three factors. First, the number of the states involved is extremely large. Every base pair can be separated or unseparated, so specification of a state of a molecule containing N base pairs involves making N binary decisions. This yields a total of 2^N distinct states of strand separation. This precludes the use of exact methods, in which all states are enumerated, to analyze molecules of biological interest, as these commonly have lengths exceeding 1,000 base pairs. Most DNAs have sites whose local sequences permit transitions to other conformations in addition to separation, further increasing the number of conformational states. Second, because the free energy needed to transform a base pair to an alternative conformation depends on the identity of the base pair involved, the analysis of equilibria must examine the specific sequence of bases in the molecule. This precludes several possible strategies for performing approximate analyses, including combinatorial methods that assume transition energetics to be the same for all base pairs, or that average the base composition of blocks. Third, and most importantly, the global and topological character of the superhelical constraint means that the conformations of all base pairs in the molecule are coupled together. Separation of a particular base pair alters its helicity, which changes the distribution of Tw, and hence of α, throughout the domain. This in turn affects the probability of transition of every other base pair. Whether transition occurs at a particular site depends not just on its local sequence, but also on how effectively this transition competes with all other alternatives. Thus, separations at particular sites can be analyzed only in the context of the entire molecule. Divide-and-conquer strategies, in which the sequence is partitioned into blocks that are individually analyzed, are thus not feasible. Superhelical transitions must be analyzed as global events, including simultaneous competitions among all possible transitions. This renders the accurate analysis of superhelical transitions extremely difficult.

It is not feasible to perform exact analyses of all states for the kilobase-length, topologically constrained molecules of biological importance, because the number of states grows exponentially. On the other hand, it is not enough to look only at the lowest energy states. Confining attention to the minimum-energy state provides a very poor depiction of transition behavior. Although any individual high-energy state is exponentially less populated, there are so many high-energy states that cumulatively they can dominate the minimum-energy state. The development of accurate methods to treat superhelical strand separation requires an intermediate approach (Benham, 1990). First, enough low-energy states must be treated exactly to provide an accurate depiction of the transition. Then the cumulative influence of the neglected, high-energy states must be estimated. Wherever possible, computed parameter values must be refined by the insertion of correction terms that account for the approximate influence of the neglected states. This is the strategy we adopt below.

THE ENERGETICS OF A STATE

A superhelical linking difference α imposed on a DNA molecule can be accommodated by three types of deformation, each of which requires free energy. First, strand separations can occur. Second, the single strands in the separated regions can twist around each other, thereby absorbing some of the linking difference. Third, the portion of α not accommodated by these alterations imposes superhelical deformations on the balance of the molecule.

Each of these deformations requires free energy that can be described by some simple formulas. Opening each new region of strand separation requires a free energy a, while separating each individual base pair within a region takes free energy b_{AT} or b_{GC}, depending on the identity of the base pair. In practice, $a \gg b_{\mathrm{GC}} > b_{\mathrm{AT}}$. Because the initiation free energy a is large, low-energy states tend to have only a small number of runs of strand separation. Because b_{GC} is larger than b_{AT}, these runs tend to be in A+T-rich regions. The free energy of interstrand twisting within separated regions is quadratic in the local helicity of the deformation, with coefficient denoted by C. The free energy of residual superhelicity has been measured experimentally to be quadratic in that deformation, with coefficient K. Combining these contributions (and allowing the interstrand twisting to equilibrate with the residual superhelicity), the free energy G of a state is found to depend on three parameters: the number n of separated base pairs, the number n_{AT} of these that are A·Ts, and the number r of runs of separation:

$$G(n, n_{\mathrm{AT}}, r) = \frac{2\pi^2 CK}{4\pi^2 C + Kn}\left(\alpha + \frac{n}{10.5}\right)^2 + ar + b_{\mathrm{AT}} n_{\mathrm{AT}} + b_{\mathrm{GC}}(n - n_{\mathrm{AT}}). \tag{7.6}$$

The energy parameters in this expression, a, b_{AT}, b_{GC}, C, and K, all depend on environmental conditions such as salt concentration and temperature. The values of the b's are known experimentally under a wide variety of conditions (Marmur and Doty, 1962; Schildkraut and Lifson, 1968). However, values for the other parameters are not so well understood. These parameters must be evaluated before the methods can yield quantitatively accurate results. We will do this by fitting these parameters to actual experimental data.

ANALYSIS OF SUPERHELICAL EQUILIBRIA

To calculate the equilibrium strand separation behavior of superhelical DNA molecules, we proceed as follows. First, the DNA sequence is analyzed and key information needed for later stages is stored. This step need be done only once per sequence. Next, the linking difference α and environmental conditions are specified, which sets the energy parameters and determines the free energy associated with each state. The state having minimum free energy under the given conditions is found from the free energy expression and the sequence data. Then an energy threshold θ is specified, and all states i are found that have free energy exceeding the minimum $G_{\mathbf{min}}$ by no more than this threshold amount. Three inequalities occur, one each for n, n_{AT}, and r. Together the satisfaction of all three inequalities provides necessary and sufficient conditions that a state satisfy the energy threshold condition. For every set of values n, n_{AT}, and r satisfying these inequalities, all states with these values are found from the sequence information. This is a very complex computational task. The number of states involved grows approximately exponentially with the threshold θ. In cases where $r > 1$, care must be taken to verify that a collection of r runs having the requisite total length and A+T-richness neither overlap nor abut, but rather are distinct. An approximate partition function $Z_{\mathbf{cal}}$ is computed from this collection of low-energy states to be

$$Z_{\mathrm{cal}} = \sum_{i:G_i - G_{\mathrm{min}} < \theta} \exp(-G_i / RT). \tag{7.7}$$

By focusing only on the low-energy states, approximate ensemble average (that is, equilibrium) values are computed for all parameters of interest. These may include the expected torsional deformation of the strand-separated regions, expected numbers of separated base pairs, of separated A·T pairs, and of runs of separation, the ensemble average free energy $\overline{G}$, and the residual superhelicity.

The most informative quantities regarding the behavior of the molecule are its destabilization and transition profiles. The transition profile displays the probability of separation of each base pair in the molecule. The separation probability $p(x)$ of the base pair at position x is calculated from equation (7.5) using parameter ζ_x, where $\zeta_x = 1$ in states where base pair x is separated and $\zeta_x = 0$ in all other states. This calculation is performed for every base pair in the sequence. The transition profile displays $p(x)$ as a function of x.

The destabilization profile is the incremental free energy needed to induce separation at each base pair. To calculate this quantity, let $i(x)$ index the states in which the base pair at position x is separated. Then the average free energy of all such states is

$$\hat{G}(x) = \frac{\sum_{i(x)} G_{i(x)} \exp(-G_{i(x)} / RT)}{\sum_{i(x)} \exp(-G_{i(x)} / RT)}. \tag{7.8}$$

To determine the destabilization free energy $G(x)$, we normalize by subtracting the calculated equilibrium free energy:

$$G(x) = \hat{G}(x) - \overline{G}. \tag{7.9}$$

Base pairs that require incremental free energy to separate at equilibrium have $G(x) > 0$, while base pairs that are energetically favored to separate at equilibrium have $G(x) \leq 0$. This calculation is performed for each base pair in the molecule, and the destabilization profile plots $G(x)$ versus x. Examples of these profiles are given in Figure 7.1.

Although individual high-energy states are exponentially less populated than low-energy states at equilibrium, they are so numerous that their cumulative contribution to the equilibrium still may be significant. The next step in this calculation requires estimating the aggregate influence of the states that were excluded from the above analysis because their free energies exceeded the threshold. This involves estimating the contribution $Z(n,n_{\mathrm{AT}},r)$ to the partition function from all states whose values n, n_{AT}, and r do not satisfy the threshold condition. Here

$$Z(n,n_{\mathrm{AT}},r) = p_{n,r}(n_{\mathrm{AT}}) M(n,r) \exp(-G(n,n_{\mathrm{AT}},r) / RT), \tag{7.10}$$

where $M(n,r)$ is the number of states with n separated base pairs in r runs, which for a circular domain is

$$M(n,r) = \frac{N}{r} \binom{n-1}{r-1} \binom{N-n-1}{r-1}. \tag{7.11}$$

The only part of $Z(n,n_{\mathrm{AT}},r)$ not amenable to exact determination is $p_{n,r}(n_{\mathrm{AT}})$, the fraction of (n,r)-states that have exactly n_{AT} separated A·T base pairs.

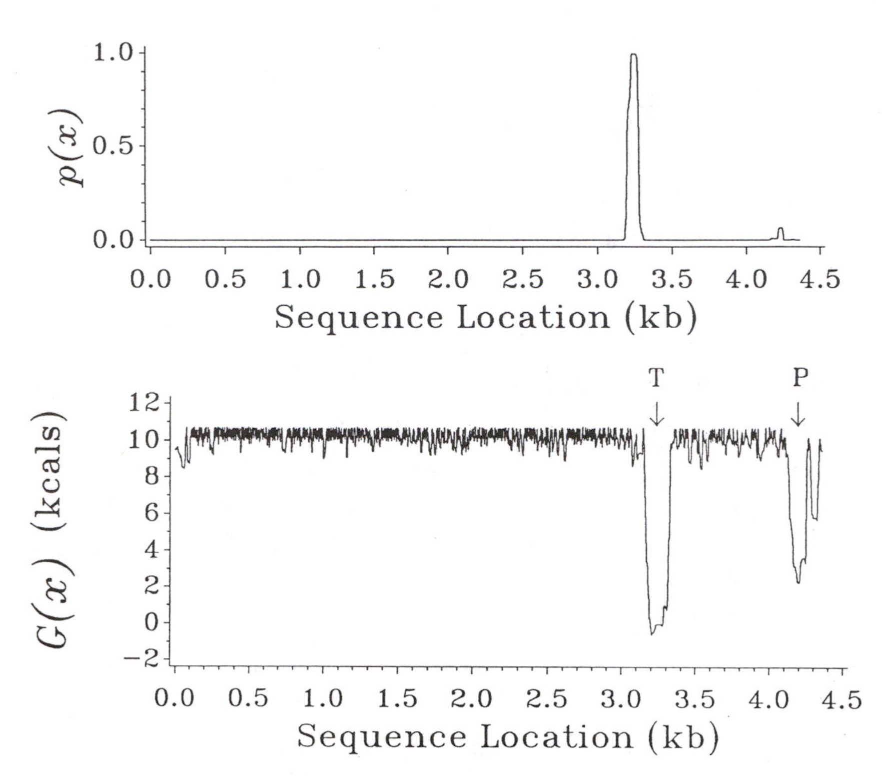

FIGURE 7.1 The transition (top) and the helix destabilization (bottom) profiles are shown for the circular pBR322 DNA molecule at $\alpha = -26$ turns. The promoter (P) and terminator (T) (the sites that control the start and end, respectively, of gene expression) of the beta-lactamase gene are indicated. These results were calculated using the energetics found by the method described in the text. Reprinted (bottom panel), by permission, from Benham (1993). Copyright © 1993 by the National Academy of Sciences.

The estimation of the influence of the high-energy states is done in two steps. First, growth conditions are found for $Z(n,r)$, the contribution to the partition function from all states having n total separated base pairs in r runs. Assuming that the distribution of A+T-richness among $(n,r)-$states is approximately the same as that for $(n,r+1)$ -states, then the ratio

$$Z(n,r+1) / Z(n,r) \approx \frac{(n-r)(N-n-r)}{r(r+1)} \exp(-a / RT) = T(r) \quad (7.12)$$

is monotonically decreasing with r. One can find an $\hat{r}$ such that $T(\hat{r}) = \rho < 1$. Then for all $r \geq \hat{r}$ we have $T(r) \leq \rho$, so that

$$\sum_{r=\hat{r}}^{r_{\max}} Z(n,r) \leq \frac{Z(n,\hat{r})}{1-\rho}. \tag{7.13}$$

By a similar line of reasoning one also can find a value $\hat{n}$, above which the aggregate contribution to the partition function again is bounded above by a convergent geometric series:

$$\sum_{n=\hat{n}}^{N-r} Z(n,r) \leq \frac{Z(\hat{n},r)}{1-\sigma}. \tag{7.14}$$

In practice, low values (≈ 0.1) for the series ratios ρ and σ occur at reasonably small cutoffs ($\hat{r} \approx 8$, $\hat{n} \approx 150$ for a molecule of $N = 5{,}000$ base pairs under reasonable environmental conditions).

The contribution of the intermediate states having $n \leq \hat{n}$ and $r \leq \hat{r}$ but not satisfying the threshold requires estimating $p_{n,r}(n_{\mathtt{AT}})$, the fraction of (n,r)-states that have exactly $n_{\mathtt{AT}}$ separated A·T base pairs. Although in principle one can compute this quantity from the base sequence, for molecules of kilobase lengths it is feasible to compute only the exact distribution of A+T-richness in $r = 2$ run states having $n \leq n_{\mathbf{max}} \approx 200$. Experience has shown that high accuracy is obtained by calculating $p_{n,2}(n_{\mathtt{AT}})$ exactly, and using $p_{n,2}(n_{\mathtt{AT}})$ as an estimate of $p_{n,r}(n_{\mathtt{AT}})$ for $r > 2$. Once the sequence information needed in this step has been found (a calculation that need be performed only once per molecule), the performance of the rest of this refinement is computationally very fast.

These results are used to estimate the contribution $\hat{Z}_{\mathbf{neg}}$ to the partition function from the neglected, high-energy states:

$$\hat{Z} = Z_{\mathbf{cal}} + \hat{Z}_{\mathbf{neg}}. \tag{7.15}$$

(Here the carat marks denote approximate values.) Any parameter ζ that depends only on n, $n_{\mathtt{AT}}$, or r also can have its previously calculated approximate equilibrium value corrected for the estimated effects of all neglected, high-energy states:

$$\hat{\zeta} = \frac{\sum_{i:G_i<G_{\min}+\theta} \zeta_i \exp(-G_i / RT) + \sum_{\text{neg}} \zeta(n, n_{\text{AT}}, r)\hat{Z}(n, n_{\text{AT}}, r)}{Z_{\text{cal}} + \hat{Z}_{\text{neg}}}. \tag{7.16}$$

Examples of correctable parameters include the population-averaged values of the total numbers of separated base pairs, runs of transition, and separated A·T pairs. The only important quantities that cannot be refined in this way are the transition and destabilization profiles, because their calculation involves positional information. However, their accuracy can be estimated by comparing the corrected ensemble average number of separated base pairs with its (uncorrected) value that is computed as the sum of the probabilities of separation for all base pairs in the sequence. In this way the accuracy of the profiles calculated with any specified threshold can be assessed. This allows the threshold to be chosen to give any required degree of accuracy. In practice accuracies exceeding 99 percent are feasible at physiological temperatures, even for highly supercoiled molecules.

Evaluation of Free-Energy Parameters

Before these techniques can yield quantitatively precise calculations, accurate values must be known for the energy parameters. Only the separation energetics b_{AT} and b_{GC} have been accurately measured under a wide range of environmental conditions (Marmur and Doty, 1962; Schildkraut and Lifson, 1968). The other parameters (the quadratic coefficient K governing residual linking, the cooperativity free energy a, and the coefficient C governing interstrand twisting of strand-separated DNA) are known only for a restricted range of molecules and environmental conditions.

The theoretical methods described above can be used to determine the best fitting values of the unknown parameters based on the analysis of experimental data on superhelical strand separation. Allowing the parameters to vary within reasonable ranges, the analyses are repeated, and the set of values is found for which the computed transition properties best fit the experimental data (Benham, 1992). Application of this method to data on strand separation in pBR322 DNA at $[Na^+] = 0.01\ \mathrm{M}$, $T = 310\ \mathrm{K}$ finds a unique optimum fit when $K = 2350 \pm 80\ RT/N$, $a = 10.84 \pm 0.2\ \mathrm{kcal}$, and $C = 2.5 \pm 0.3 \times 10^{-13}\ \text{erg-nt/rad}^2$.

Extensive sample calculations of strand separations in superhelical DNA have been performed using these energy parameters (Benham, 1992). As described above, substantial amounts of free energy are required to drive strand separation. In consequence, this transition is favored only when the DNA is significantly supercoiled. This is shown in Figure 7.2, where the solid line depicts the probability of strand separation in pBR322 DNA (N = 4,363 base pairs) as a function of imposed negative superhelicity under low-salt conditions. The dashed line gives the ensemble average number of strand-separated base pairs as a function of $-\alpha$. Separation occurs only when the linking difference satisfies $\alpha \leq -18$ turns and is confined to the terminator (3,200 to 3,300) and promoter (4,100 to 4,200) regions of one particular gene, as shown in Figure 7.1 above. These results are in precise agreement with experiment (Kowalski et al., 1988).

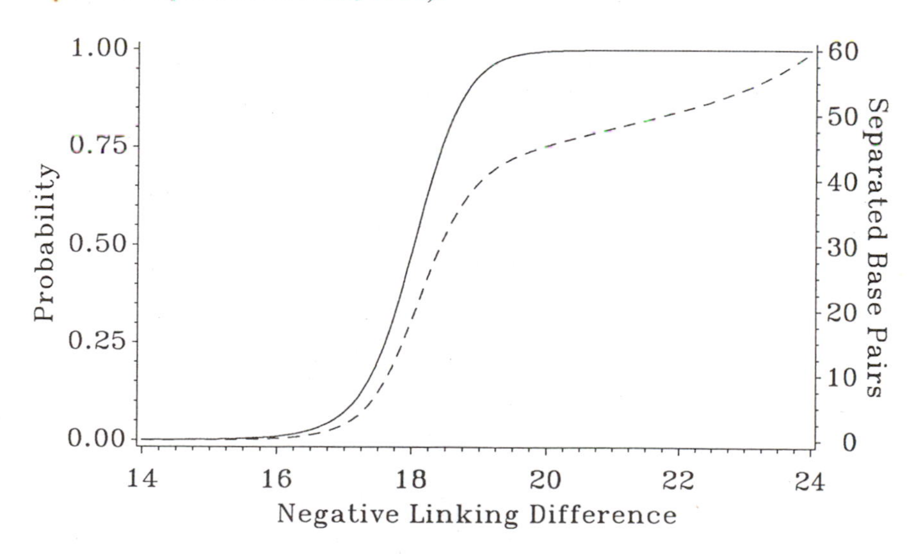

FIGURE 7.2 The onset of strand separation in pBR322 DNA.

Accuracy of the Calculated Results

Once the energetics governing transition under specific environmental conditions have been fit based on the transition behavior of one sequence, the accuracy of the analytical methods can be assessed by comparing their predictions with experimental results on other molecules (Benham, 1992). We did this for six DNA molecules synthesized by David Kowalski, a biochemist studying the role of strand separation in initiating replication (Kowalski and Eddy, 1989). Starting from a parent DNA molecule pORIC, Kowalski made various modifications. pDEL16 has a 16-base pair deletion from the replication origin site of pORIC. pAT105 and pGC91 were made by inserting an A+T-rich 105-base pair segment and a G+C-rich 91-base pair segment, respectively, into the deletion site of pDEL16. pAT105I and pGC91I have the same insertions, but placed in reverse orientation. The complete DNA sequences of these plasmids were provided to the author by Dr. Kowalski (private communication).

The transition profiles of these molecules were calculated using the energetics appropriate to the experimental conditions, which were the same as in the pBR322 experiments from which the energy parameters were derived. Figure 7.3 shows the computed transition profiles around the duplex unwinding element (DUE) of the origin site for the four plasmids of greatest interest. The region where strand separation was detected experimentally is shown by a double line in each case. Less separation was detected experimentally at this location in the pORIC plasmid than in the other two transforming molecules, and none was detected in pDEL16 or in the other two molecules whose profiles are not shown in the figure. These experimental results are in close agreement with the present predictions. In fact, the agreement may be even better than the figure indicates. Because the experimental method detects separation only in the interiors of open regions, the actual separated sites are slightly larger than what the experiment detects.

These results show that the present methods for analyzing superhelical strand separation are highly accurate. The extensive variations in the locations of separated regions that result from minor sequence alterations are precisely depicted. The relative amounts of transition at each site also agree closely with experiment. The superhelicity required to drive a specific amount of separation is within 7 percent of the observed value, which reflects the limit of accuracy with which extents of transition are

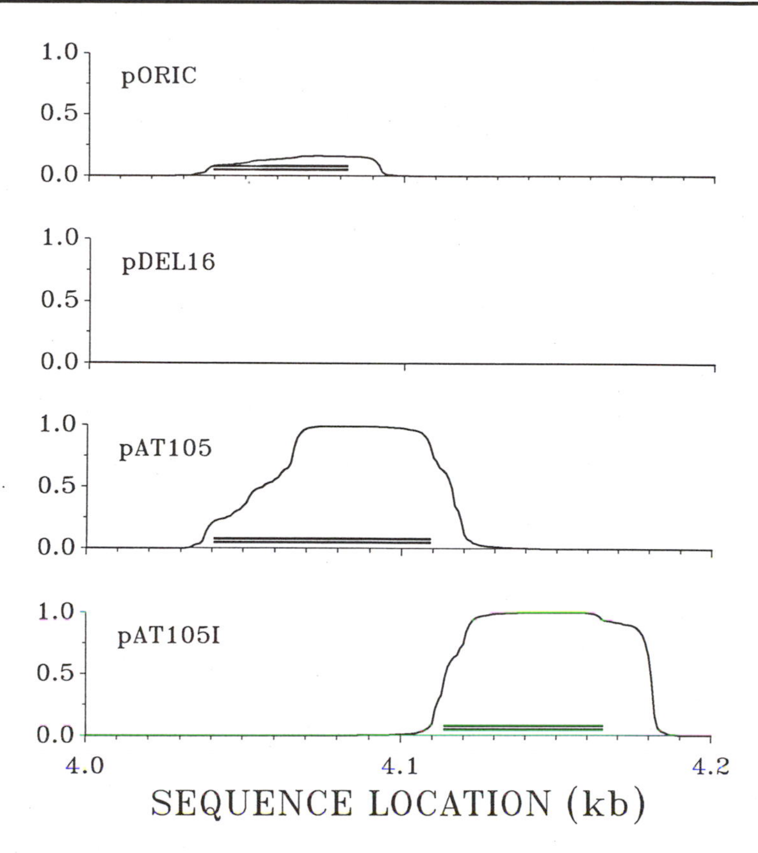

FIGURE 7.3 The transition profiles at DUE sequences. Reprinted, by permission, from Benham (1992). Copyright © 1992 by Academic Press Limited.

measured in these experiments. This demonstrates that these analytical methods provide highly precise predictions of the details of strand separation in superhelical molecules.

APPLYING THE METHOD TO STUDY INTERESTING GENES

Having developed a method and confirmed its accuracy on test molecules, we can now apply it to study any DNA sequence of interest. It turns out to be particularly illuminating to examine the association between sites of superhelical destabilization and sites of gene regulation

(Benham, 1993). Our calculations show some striking correlations involving sites for initiation of transcription, termination of transcription, initiation of DNA replication, and binding of repressor proteins.

We find that some bacterial genes show superhelical destabilization at the sites where gene expression starts and the sites where it ends. One gene on the pBR322 DNA molecule (from which the data in Figure 7.1 came) and one on the ColE1 plasmid (from which the data in Figure 7.4a came) are bracketed by such sites, suggesting that their expression is regulated by the state of DNA supercoiling. And, indeed, experiments show that these bracketed genes are expressed at higher rates when their DNA is superhelical than when it is relaxed. The other genes on these molecules show no such destabilized regions. This result suggests that genes in bacterial DNA can be partitioned into two categories, depending on whether or not they are bracketed by superhelically destabilized regulatory regions.

In a similar vein, we have analyzed the DNA sequences of two mammalian viruses, the polyoma and papilloma viruses, each of which can cause cancer. The most destabilized locations on these molecules occur precisely at the places where gene expression terminates, the so-called poly-adenylation sites. The two most destabilized sites in the polyoma genome occur at the major (M) and minor (m) poly-adenylation sites, as shown in Figure 7.4b. Of the three most destabilized sites in the papilloma virus genome, two occur at known poly-adenylation sites for transcription from the direct strand. The other occurs at a location having the sequence attributes of a poly-adenylation site for transcription from the complementary strand. (This observation raises the intriguing possibility that the complementary strand of this molecule could transcribe, an event that has not been observed to date.)

The strong association found between destabilized sites and the beginnings and ends of genes suggests that destabilization may play roles in their functioning. Many possible scenarios can be suggested for how this could occur. Clearly, destabilization at a gene promoter could facilitate the start of transcription by assisting the formation of a complex between the single strand to be transcribed and the enzyme complex that constructs the RNA transcript. What about destabilization at the sites where gene expression is completed (terminators in bacteria and poly-adenylation sites in higher organisms)? In this case, a likely but subtler role can be suggested. The moving transcription apparatus is thought to

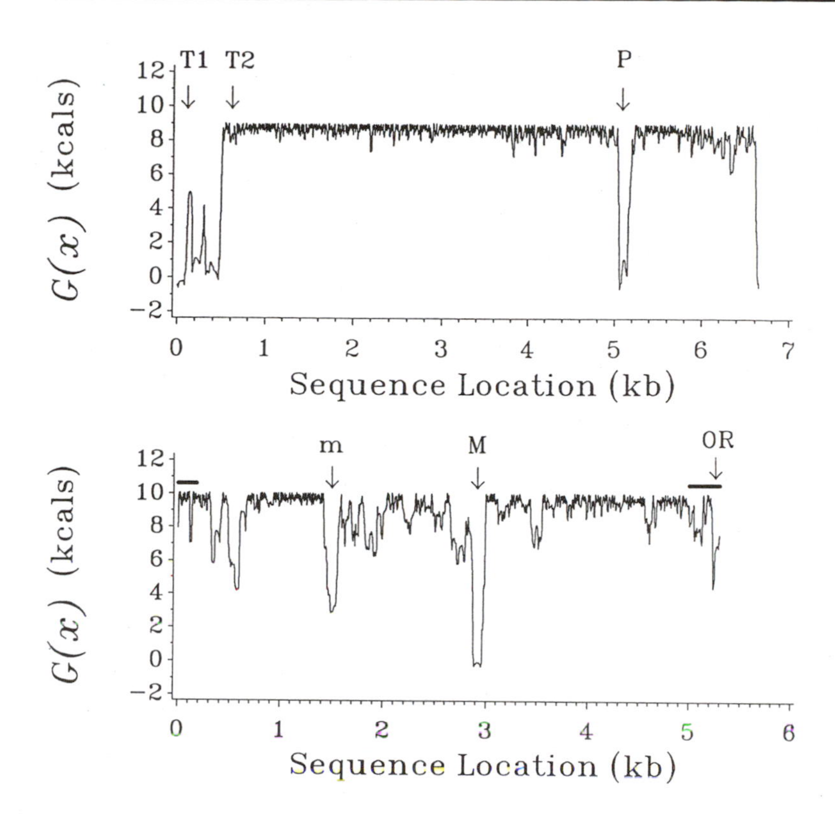

FIGURE 7.4 The helix destabilization profiles of the circular molecules (top) ColE1 plasmid DNA and (bottom) polyoma virus DNA. The locations of the promoter (P) and terminators (T1 and T2) of the bracketed transcription unit of ColE1 are indicated. In polyoma the control region (denoted by a bar), replication origin (OR), and the major (M) and minor (m) poly-adenylation sites are shown. Reprinted, by permission, from Benham (1993). Copyright © 1993 by the National Academy of Sciences.

push a wave of positive supercoils ahead and leave a wake of negative supercoils behind (Wu et al., 1988). A region of strand separation constitutes a localized concentration of negative superhelicity, due to the large decrease in twist that occurs. This could provide a sink for the positive supercoils generated by an approaching complex, preventing the accumulation of twisting and bending deformations that otherwise could impede its progress. This would facilitate efficient transcription of the gene involved. The wake of negative supercoils left behind could destabilize the promoter region in preparation for the next round of expression. This model could explain why terminal regions are the most destabilized sites found, and why some genes are bracketed by destabilized sites.

DNA replication is another process for which it is interesting to study the correlation with superhelical destabilization. In the plasmids pBR322 and ColE1, replication is started by an RNA primer, which displaces one strand of the DNA at the replication origin by base pairing to the strand having a complementary sequence. The origin sites on these molecules are not destabilized by superhelicity, suggesting that the displacement event does not require a destabilized or separated site. By contrast, replication of the DNA of phage f1 (a virus that attacks bacteria) involves enzymatic cutting of one strand that is known to require DNA superhelicity. If one role of DNA superhelicity is to promote strand separation, one would expect to find highly destabilized sites abutting the origin of replication. In fact, the calculations show precisely this, providing strong support for the assumption.

Interesting results also emerge from the study of genes involved in the SOS response system in the bacterium *E. coli*. The SOS response system, as its name suggests, is a collection of genes that are turned on when the organism experiences any of a variety of serious problems, ranging from environmental stresses to DNA damage. These genes are usually turned off by the binding to their promoters of a repressor protein called LexA, which blocks transcription. Another protein, called RecA, plays a key role in initiating the SOS response by causing the removal of the LexA repressor, thereby allowing transcription.

As it happens, the ColE1-encoded gene discussed above that was bracketed by superhelical destabilization sites is a member of the SOS response system. What about other SOS response genes? Do they also show superhelically destabilized regions? To address this question, we examined every known SOS response gene whose DNA sequence was available. In every case, the LexA binding site was contained in a strongly destabilized region. We note that this binding site is 16 base pairs long. Although it is reasonably A+T-rich, this is not long enough for its presence alone to assure destabilization.

It is not hard to speculate on the function of these superhelical destabilization regions in the SOS response. The RecA protein is known to bind single-stranded DNA. If the SOS response is marshaled against DNA damage, the damaged region provides single-stranded DNA for RecA binding. The environmental stresses that activate the SOS response system cause an increase in DNA superhelicity (Bhriain et al., 1989), which can provide strand-separated regions near the LexA binding sites that allow RecA to bind.

In addition to those sketched here, many other possible roles for superhelically destabilized regions can be suggested. For example, destabilization of one site in a molecule could protect other sites from separating that must remain in the duplex form to function.

DISCUSSION AND OPEN PROBLEMS

This chapter has described how DNA sequences can be analyzed to determine one biologically important attribute—the relative susceptibility of regions in the molecule to superhelical destabilization. The last section indicated how correlations between destabilized sites and DNA regulatory regions illuminate the mechanisms of activity of such regions. This work has many other possible uses, one of which is sketched here. Correlations of the type noted above between superhelically destabilized sites and regulatory regions can be used in searching DNA sequences for those regions.

Most commonly available strategies search DNA sequences for short subsequences (that is, strings) whose presence correlates with a particular activity. So-called `TATA` boxes are present at promoters, for example, while poly-adenylation occurs near `AATAAA` sites. Sequence signatures are known for terminators and for several other types of regulatory sites. This string-search approach is possible because the enzymes involved with particular functions usually have either specific or consensus sequence requirements for activity. In most cases these string-search methods find large numbers of candidate sites having the sequence characteristics necessary for function. Among these, commonly only a small number of sites actually are active.

The strong associations documented here between destabilized sites and particular types of regulatory regions suggest that this attribute also could be used to search genomic sequences for those regions. This would supplement existing string methods, providing more accurate predictions. For example, the bovine papilloma virus DNA sequence contains 9 sites having the `AATAAA` sequence needed for poly-adenylation, of which only 2 are known to be active. The most destabilized sites on the molecule contain 6 of these signal sequences, several of which are very close together, including both known poly-adenylation sites. As a second example, a search of all known *E. coli* sequences finds more than 100 locations having the sequence associated with LexA binding. Analysis of which of these sites are destabilized could suggest whether some might be promoters for previously unrecognized SOS-regulated genes.

The transition behavior of stressed DNA molecules can be complicated by several additional factors. First, there are other types of transitions possible for specific sequences within a DNA molecule. For example, sequences in which a purine (`A` or `G`) alternates with a pyrimidine (`C` or `T`) along each strand can adopt a left-handed helical structure. Transitions to this and to other alternative conformations also can be driven by imposed superhelicity. So the equilibrium experienced by a stressed molecule actually involves competition among several types of transitions, not just strand separation. Because these other conformations usually are possible only at a small number of short sites having the correct sequence, their analysis is combinatorially simpler than the treatment of strand separation. The theoretical methods described here are currently being extended to include the possible occurrence of other types of transitions.

The second complication arises from the structural restraints on DNA in cells. There the DNA is not free to twist and writhe to minimize its energy, but instead is wound around basic proteins to form a chromatin fiber. This drastically alters the types of deformations the molecule can undergo. While it is not clear precisely how this constraint interacts with superhelicity, conformational transitions are expected to be driven by less extreme deformations in restrained molecules than in unrestrained ones (Benham, 1987).

The approach outlined here has great promise for finding biologically important correlates of regulation and for illuminating specific mechanisms of function.

REFERENCES

Benham, C.J., 1987, "The influence of tertiary structural restraints on conformational transitions in superhelical DNA," *Nucleic Acids Res.* **15**, 9985-9995.

Benham, C.J., 1990, "Theoretical analysis of heteropolymeric transitions in superhelical DNA molecules of specified sequence," *J. Chem. Phys.* **92**, 6294-6305.

Benham, C.J., 1992, "Energetics of the strand separation transition in superhelical DNA," *Journal of Molecular Biology* **225**, 835-847.

Benham, C.J., 1993, "Sites of predicted stress-induced DNA duplex destabilization occur preferentially at regulatory loci," *Proceedings of the National Academy of Sciences USA* **90**, 2999-3003.

Bhriain, N. Ni, C. Dorman, and C. Higgins, 1989, "An overlap between osmotic and anaerobic stress responses: A potential role for DNA supercoiling in the coordinate regulation of gene expression," *Mol. Microbiol.* **3**, 933-942.

Dorman, C., G. Barr, N. Ni Bhriain, and C.F. Higgins, 1988, "DNA supercoiling and the anaerobic and growth phase regulation of *tonB* gene expression," *J. Bacteriol.* **170**, 2816-2826.

Gellert, M., 1981, "DNA topoisomerases," *Annu. Rev. Biochem.* **50**, 879-910.

Hartwig, M., E. Matthes, and W. Arnold, 1981, "Extremely underwound chromosomal DNA in nucleoids of mouse sarcoma cells," *Cancer Letters* **13**, 153-158.

Kowalski, D., and M. Eddy, 1989, "The DNA unwinding element: A novel *cis*-acting component that facilitates opening of the *Escherichia coli* replication origin," *EMBO J.* **8**, 4335-4344.

Kowalski, D., D. Natale, and M. Eddy, 1988, "Stable DNA unwinding, not breathing, accounts for single-strand specific nuclease hypersensitivity of specific A+T-rich regions," *Proceedings of the National Academy of Sciences USA* **85**, 9464-9468.

Malkhosyan, S., Y. Panchenko, and A. Rekesh, 1991, "A physiological role for DNA supercoiling in the anaerobic regulation of colicin gene expression," *Mol. Gen. Genet.* **225**, 342-345.

Marmur, J., and P. Doty, 1962, "Determination of the base composition of deoxyribonucleic acid from its thermal denaturation temperature," *Journal of Molecular Biology* **5,** 109-118.

Mattern, M., and R. Painter, 1979, "Dependence of mammalian DNA replication on DNA supercoiling," *Biochim. Biophys. Acta* **563**, 293-305.

Pruss, G., and K. Drlica, 1989, "DNA supercoiling and prokaryotic transcription," *Cell* **56**, 521-523.

Schildkraut, C., and S. Lifson, 1968, "Dependence of the melting temperature of DNA on salt concentration," *Biopolymers* **3**, 195-208.

Smith, G., 1981, "DNA supercoiling: Another level for regulating gene expression," *Cell* **24**, 599-600.

Umek, R., M. Linskens, D. Kowalski, and J. Huberman, 1989, "New beginnings in studies of eucaryotic DNA replication origins," *Biochem. Biophys. Acta* **1007,** 1-14.

Weintraub, H., P. Cheng, and K. Conrad, 1986, "Expression of transfected DNA depends on DNA topology," *Cell* **46,** 115-122.

White, J.H., 1988, "An introduction to the geometry and topology of DNA structure," pp. 225-254 in *Mathematical Methods for DNA Sequences,* M.S. Waterman (ed.), Boca Raton, Fla.: CRC Press.

Wu, H., S. Shyy, J.C. Wang, and L.F. Liu, 1988, "Transcription generates positively and negatively supercoiled domains in the template," *Cell* **53**, 433-440.

Chapter 8
Lifting the Curtain: Using Topology to Probe the Hidden Action of Enzymes

De Witt Sumners
Florida State University

A central problem in molecular biology is understanding the mechanism by which enzymes carry out chemical transformations. The problem is challenging because most experimental techniques provide only a static snapshot, not a moving picture, of the sequence of molecular events that take place inside the catalytic core of the enzyme. For one class of enzymes, however, mathematics provides a powerful tool to the molecular biologist. These enzymes are the ones that perform topological reactions necessary for the winding, unwinding, recombination, and transposition of DNA. Using topological results about knots and tangles, one can peer into the reaction center and infer the mechanisms of action.

One of the important issues in molecular biology is the three-dimensional structure (shape) of proteins and deoxyribonucleic acid (DNA) in solution in the cell, and the relationship between structure and function. Ordinarily, protein and DNA structure is determined by X-ray crystallography or electron microscopy. Because of the close packing needed for crystallization and the manipulation required to prepare a specimen for electron microscopy, these methods provide little direct evidence for molecular shape in solution. The three-dimensional shape in solution is of great biological significance but is very difficult to determine (Wang, 1982).

Experimental techniques such as X-ray crystallography and nuclear magnetic resonance provide ways to infer precise distances between atoms. However, these methods are not well suited to studying the dynamic mechanism by which enzymes act. Interestingly, topology can shed light on this key issue. The topological approach to enzymology is an experimental protocol in which the descriptive and analytical powers of topology and geometry are employed in an indirect effort to determine the enzyme mechanism and the structure of active enzyme-DNA complexes in vitro (in a test tube) (Wasserman and Cozzarelli, 1986; Sumners, 1987a). Once the enzyme structure and mechanism are understood in a controlled laboratory situation, this knowledge can be extrapolated to enzyme mechanism in vivo, that is, in a living cell.

Topology is a branch of mathematics related to geometry. It is often characterized as "rubber-sheet geometry," because topological equivalence of spaces allows stretching, shrinking, and twisting of an object in order to make it congruent to another object. Topology is the study of properties of objects (spaces) that are unchanged by allowable elastic deformations. When a given topological property differs for a pair of spaces, then one can be sure that one space cannot be transformed into the other by elastic deformation. Changes that can produce non-equivalent spaces include cutting the space apart and reassembling the parts to produce another space. It is precisely this topological breakage and reassembly of DNA that characterizes the mechanism of some life-sustaining cellular enzymes, enzymes that facilitate replication, transcription, and transposition. Chapter 6 describes aspects of the geometry and topology of DNA and points out various topological transformations that must be performed on DNA by enzymes in order to carry out the life cycle of the cell. In the present chapter, we describe how recent results in three-dimensional topology (Culler et al., 1987; Ernst and Sumners, 1990; Sumners, 1990, 1992) have proven to be of use in the description and quantization of the action of these life-sustaining enzymes on DNA.

THE TOPOLOGY OF DNA

The DNA of all organisms has a complex and fascinating topology. It can be viewed as two very long curves that are intertwined millions of times, linked to other curves, and subjected to four or five successive orders of coiling to convert it into a compact form for information storage. If one scales the cell nucleus up to the size of a basketball, the DNA inside scales up to the size of thin fishing line, and 200 km of that fishing line are inside the nuclear basketball. Most cellular DNA is double-stranded (duplex), consisting of two linear backbones of alternating sugar and phosphorus. Attached to each sugar molecule is one of the four bases (nucleotides): A = adenine, T = thymine, C = cytosine, G = guanine. A ladder whose sides are the backbones and whose rungs are hydrogen bonds is formed by hydrogen bonding between base pairs, with A bonding only with T, and C bonding only with G. The base pair sequence for a linear segment of duplex DNA is obtained by reading along one of the two backbones, and is a word in the letters {A,T,C,G}. Due to the uniqueness of the bonding partner for each nucleotide, knowledge of the sequence along one backbone implies knowledge of the sequence along the other backbone. In the classical Crick-Watson double helix model for DNA, the ladder is twisted in a right-hand helical fashion, with an average and nearly constant pitch of approximately 10.5 base pairs per full helical twist. The local helical pitch of duplex DNA is a function of both the local base pair sequence and the cellular environment in which the DNA lives; if a DNA molecule is under stress, or constrained to live on the surface of a protein, or is being acted upon by an enzyme, the helical pitch can change. Duplex DNA can exist in nature in closed circular form, where the rungs of the ladder lie on a twisted cylinder. Circular duplex DNA exists in the mitochondria of human cells, for example. Duplex DNA in the cell nucleus is a linear molecule, one that is topologically constrained by periodic attachment to a protein scaffold in order to achieve efficient packing.

The packing, twisting, and topological constraints all taken together mean that topological entanglement poses serious functional problems for DNA. This entanglement would interfere with, and be exacerbated by, the vital life processes of replication, transcription, and recombination (Cozzarelli, 1992). For information retrieval and cell viability, some geometric and topological features must be introduced into the DNA, and others quickly removed (Wang, 1982, 1985). For example, the Crick-Watson helical twist of duplex DNA may require local unwinding in order to make room for a protein involved in transcription to attach to the DNA. The DNA sequence in the vicinity of a gene may need to be altered to include a promoter or repressor. During replication, the daughter duplex DNA molecules become entangled and must be disentangled in order for replication to proceed to completion. After introduction of these life-sustaining changes in DNA geometry and topology, and after the process that these changes make possible is finished, the original DNA conformation must be restored. Some enzymes maintain the proper geometry and topology by passing one strand of DNA through another by means of a transient enzyme-bridged break in one of the DNA strands, a move performed by topoisomerases. Other enzymes break the DNA apart and recombine the ends by exchanging them, a move performed by recombinases. The description and quantization of the three-dimensional structure of DNA and the changes in DNA structure due to the action of these enzymes have required the serious use of geometry and topology in molecular biology. Geometry and topology provide ways of inferring the dynamic process of topological transformation carried out by an enzyme. This use of mathematics as an analytic tool for the indirect determination of enzyme mechanism is especially important because there is no experimental way to observe the dynamics of enzymatic action directly.

In the experimental study of DNA structure and enzyme mechanism, biologists developed the topological approach to enzymology (Wasserman and Cozzarelli, 1986; Sumners, 1987b). In this approach, one performs experiments on circular substrate DNA molecules. These circular substrate molecules are genetically engineered by cloning techniques to contain regions that a certain enzyme will recognize and act upon. The circular form of the substrate molecule traps an enzymatic topological signature in the form of DNA knots and links (catenanes). Trapping such a topological signature is impossible if one uses linear DNA substrate. These DNA knots and links are observed by gel electrophoresis and electron microscopy of the reaction product DNA molecules. By observing the changes in geometry (supercoiling) and topology (knotting and linking) in DNA caused by an enzyme, the enzyme mechanism can be described and quantized. Figure 8.1a gives the schematics of the topological enzymology protocol; the black box represents the dynamic reaction in which the enzyme attaches to the DNA substrate, breaks it apart and reconnects as necessary, and then releases the DNA products. Typical results of this experimental protocol are the reaction products displayed in Figures 8.1b and 8.1c. Figure 8.1b shows the electron micrograph of a DNA (+) figure eight catenane (Krasnow et al., 1983), and Figure 8.1c shows a micrograph of the DNA knot 6_2* (Wasserman et al., 1985). Both are products of processive Tn3 recombination and are explained in detail below.

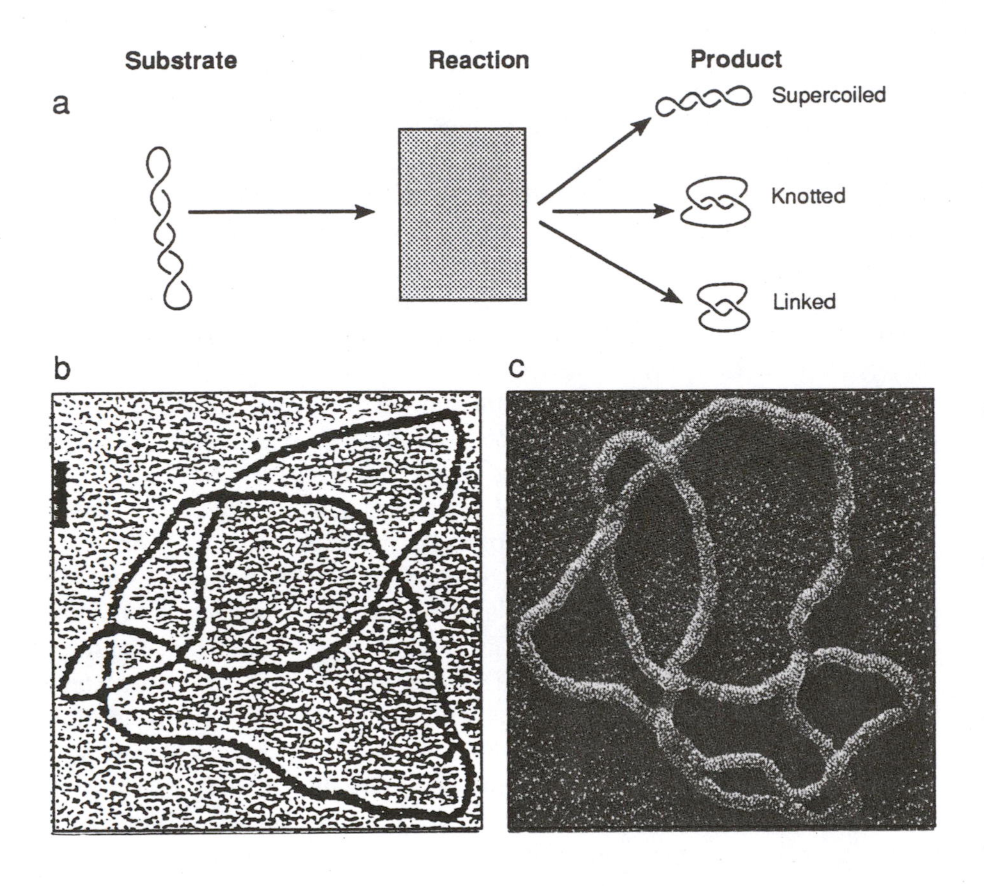

FIGURE 8.1 (a) Topological approach to enzymology. (b) DNA (+) figure eight catenane, (c) DNA knot 6_2^*. Figure 8.1b reprinted, with permission, from Krasnow et al. (1983). Copyright © 1983 by Macmillian Magazines Limited. Figure 8.1c reprinted, by permission, from Wasserman et al. (1985). Copyright © 1985 by the American Association for the Advancement of Science.

The topological approach to enzymology poses an interesting challenge for mathematics: from the observed changes in DNA geometry and topology, how can one mathematically deduce enzyme mechanisms? This requires the construction of mathematical models for enzyme action and the use of these models to analyze the results of topological enzymology experiments. The entangled form of the product DNA knots and links contains information about the enzymes that made them. Mathematics is required to extract mechanism information from the topological structure of the reaction products. In addition to utility in the analysis of experimental results, the use of mathematical models forces all of the background assumptions about the biology to be carefully laid out. At this point they can be examined and dissected and their influence on the biological conclusions drawn from experimental results can be determined.

SITE-SPECIFIC RECOMBINATION

Site-specific recombination is one of the ways in which nature alters the genetic code of an organism, either by moving a block of DNA to another position on the molecule (a move performed by transposase) (Sherratt et al., 1984) or by integrating a block of alien DNA into a host genome (a move performed by integrase). One of the biological purposes of recombination is the regulation of gene expression in the cell, because it can alter the relative position of the gene and its repressor and promoter sites on the genome. Site-specific recombination also plays a vital role in the life cycle of certain viruses, which utilize this process to insert viral DNA into the DNA of a host organism. An enzyme that mediates site-specific recombination on DNA is called a recombinase. A recombination site is a short segment of duplex DNA whose sequence is recognized by the recombinase. Site-specific recombination can occur when a pair of sites (on the same or on different DNA molecules) become juxtaposed in the presence of the recombinase. The pair of sites is aligned through enzyme manipulation or random thermal motion (or both), and both sites (and perhaps some contiguous DNA) are then bound by the enzyme. This stage of the reaction is called synapsis, and we will call this intermediate protein-DNA complex formed by the part of the substrate that is bound to the enzyme together with the enzyme itself the synaptosome (Benjamin and Cozzarelli, 1990; Heichman and Johnson, 1990; Pollock and Nash, 1983; Griffith and Nash, 1985; Kim and Landy, 1992). We will call the entire DNA molecule(s) involved in synapsis (including the parts of the DNA molecule(s) not bound to the enzyme), together with the enzyme itself, the synaptic complex. The electron micrograph in Figure 8.2 shows a synaptic complex formed by the recombination enzyme Tn3 resolvase when reacted with unknotted circular duplex DNA. In the micrograph of Figure 8.2, the synaptosome is the black mass attached to the DNA circle, with the unbound DNA in the synaptic complex forming twisted loops in the exterior of the synaptosome. It is our intent to deduce mathematically the path of the DNA in the black mass of the synaptosome, both before and after recombination. We want to answer the question: How is the DNA wound around the enzyme, and what happens during recombination?

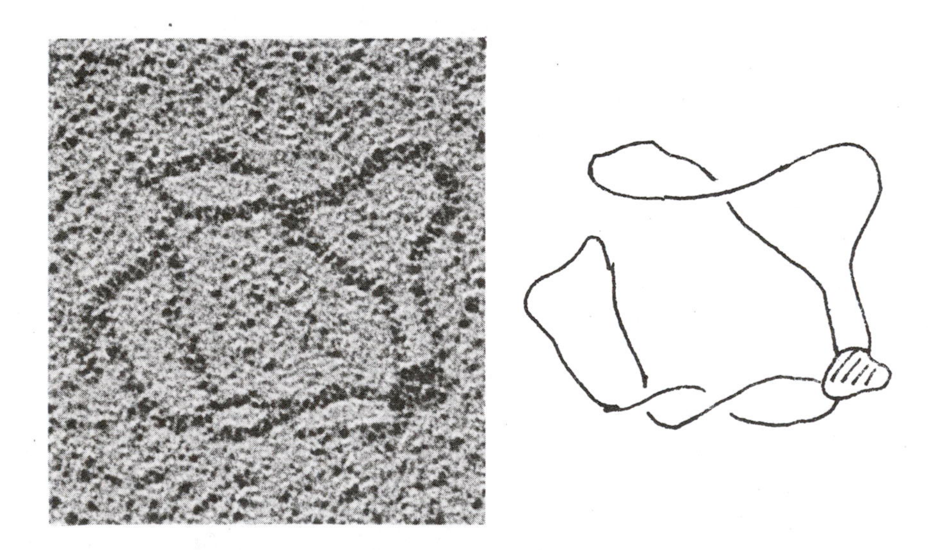

FIGURE 8.2 Tn3 synaptic complex. (Courtesy of N.R. Cozzarelli.)

After forming the synaptosome, a single recombination event occurs: the enzyme then performs two double-stranded breaks at the sites and recombines the ends by exchanging them in an enzyme-specific manner. The synaptosome then dissociates, and the DNA is released by the enzyme. We call the pre-recombination unbound DNA molecule(s) the substrate and the post-recombination unbound DNA molecule(s) the product. During a single binding encounter between enzyme and DNA, the enzyme may mediate more than one recombination event; this is called processive recombination. On the other hand, the enzyme may perform recombination in multiple binding encounters with the DNA, which is called distributive recombination. Some site-specific recombination enzymes mediate both distributive and processive recombination.

Site-specific recombination involves topological changes in the substrate. In order to identify these topological changes, one chooses to perform experiments on circular DNA substrate. One must perform an experiment on a large number of circular molecules in order to obtain an observable amount of product. Using cloning techniques, one can synthesize circular duplex DNA molecules, which contain two copies of a recombination site. At each recombination site, the base pair sequence is in general not palindromic (the base pair sequence for the site read left-to-right is different from the base pair sequence read right-to-left), and hence induces a local orientation (arrow) on the substrate DNA circle. If these induced orientations from a pair of sites on a single circular molecule agree, this site configuration is called direct repeats (or head-to-tail), and if the induced orientations disagree, this site configuration is called inverted repeats (or head-to-head). If the substrate is a single DNA circle with a single pair of directly repeated sites, the recombination product is a pair of DNA circles and can form a DNA link (or catenane) (Figure 8.3). If the substrate is a pair of DNA circles with one site each, the product is a single DNA circle (Figure 8.3 read in reverse) and can form a DNA knot (usually with direct repeats). In processive recombination on a circular substrate with direct repeats, the products of an odd number of rounds of processive recombination are DNA links, and the products of an even number of rounds of processive recombination are DNA knots. If the substrate is a single DNA circle with inverted repeats, the product is a single DNA circle and can form a DNA knot. In all figures where DNA is represented by a line drawing (such as Figure 8.3), duplex DNA is represented by a single line, and supercoiling is omitted.

The experimental strategy in the topological approach to enzymology is to observe the enzyme-caused changes in the geometry and topology of the DNA and to deduce the enzyme mechanism from these changes, as in Figure 8.1a. The geometry and topology of the circular DNA substrate are experimental control variables. The geometry and topology of the recombination reaction products are observables. In vitro experiments usually proceed as follows: Circular substrate is prepared, with all of the substrate molecules representing the same knot type (usually the unknot, that is, a curve without knots). The amount of supercoiling of the substrate molecules (the supercoiling density) is also a control variable. The substrate molecules are reacted with a high concentration of purified enzyme, and the reaction products are fractionated by gel electrophoresis. DNA molecules are naturally

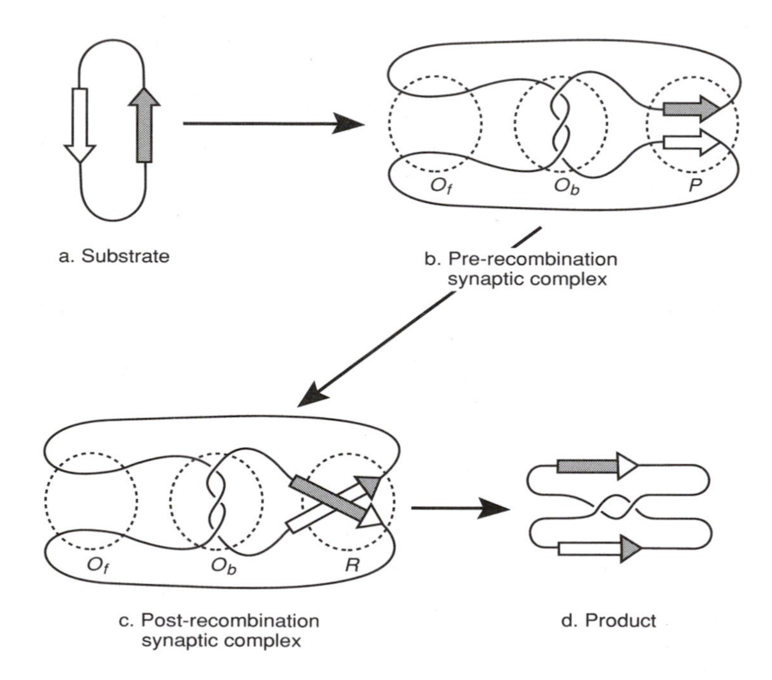

FIGURE 8.3 A single recombination event: direct repeats.

negatively charged, with the amount of negative charge proportional to the molecular weight. A gel is a resistive medium through which the DNA molecules can be forced to migrate under the influence of an electric field. The DNA sample is placed at the top of a gel column, and similar molecules migrate through the gel with similar velocities, forming discrete DNA bands in the gel when the electric field is turned off. Normally, gel electrophoresis discriminates among DNA molecules on the basis of molecular weight; given that all molecules are the same molecular weight (as is the case in these topological enzymology experiments), electrophoresis discriminates on the basis of subtle differences in the geometry (supercoiling) and topology (knot and link type) of the DNA molecules. For example, in unknotted DNA, gel electrophoresis discriminates on the basis of number of supercoils and can detect a difference of one in the number of supercoils. In gel electrophoresis of knotted and linked DNA, one must nick (break one of the two backbone strands of) the reaction products prior to electrophoresis in order to relax the supercoils in the DNA knots and links, because supercoiling confounds the gel migration of knotted and linked DNA. For nicked DNA knots and links, under the proper conditions gel velocity is (surprisingly) determined by the crossing number of the knot or link; knots and links of the same crossing number migrate with the same gel velocities (Dean et al., 1985); the higher the crossing number, the greater the gel mobility. After the gel is run, the gel bands are excised, and the DNA molecules are removed from the gel and coated with RecA protein. It is this new observation technique (RecA-enhanced electron microscopy) (Krasnow et al., 1983) that makes possible the detailed knot-theoretic analysis of reaction products. RecA is an *E. coli* protein that binds to DNA and mediates general recombination in *E. coli*. Naked (uncoated) duplex DNA is approximately 20 angstroms in diameter, and RecA-coated DNA is approximately 100 angstroms in diameter. The process of RecA coating fattens, stiffens, and stretches (untwists) the DNA. This fattening and stiffening facilitates the unambiguous determination of crossings (nodes) in an electron micrograph of a DNA knot or link and reduces the number of extraneous crossings. After RecA coating, the DNA is shadowed with platinum for viewing under the electron microscope. Electron micrographs of the reaction products (Figure 8.1b and c) are made, and frequency distributions of knot types of the products are prepared. This new precision in the determination of the topology of the reaction product spectrum opens the door for the building of detailed topological models for enzyme action.

TOPOLOGICAL TOOLS FOR DNA ANALYSIS

In this section, we will describe the parts of knot theory and tangle calculus of biological relevance. We give intuitive definitions that appeal to geometric imagination. For a rigorous mathematical treatment we refer the reader to Burde and Zieschang (1985), Kauffman (1987), and Rolfsen (1990) for knot theory and Ernst and Sumners (1990) for tangle calculus.

Knot theory is the study of the entanglement of flexible circles in 3-space. The equivalence relation between topological spaces is that of homeomorphism. A homeomorphism $h: X \to Y$ between topological spaces is a function that is one-to-one and onto, and both h and h^{-1} are continuous. An embedding of X in Y is a function $f: X \to Y$ such that f is a homeomorphism from X onto $f(X) \subset Y$. An embedding of X in Y is the placement of a copy of X into the ambient space Y. We will usually take Euclidean 3-space $\mathbf{R}^3$ (*xyz*-space) as our ambient space. A knot K is an embedding of a single circle in $\mathbf{R}^3$; a link L is an embedding of two or more circles in $\mathbf{R}^3$. For a link, each of the circles of L is called a component of L. In chemistry and biology a nontrivial link is called a catenane, from the Latin *cataena* for "chain," since the components of a catenane are topologically entangled with each other like the links in a chain. In this excursion, we will restrict attention to dimers, that is, links of two components, because dimers are the only links that turn up in topological enzymology experiments. We regard two knots (links) to be equivalent if it is possible to continuously and elastically deform one embedding (without breaking strands or passing strands one through another) until it can be superimposed upon the other. More precisely, if K_1 and K_2 denote two knots (links) in $\mathbf{R}^3$, they are equivalent (written $K_1 = K_2$) if and only if there is a homeomorphism of pairs $h: (\mathbf{R}^3, K_1) \to (\mathbf{R}^3, K_2)$ that preserves orientation on the ambient space $\mathbf{R}^3$. We take our ambient space $\mathbf{R}^3$ to have a fixed (right-handed) orientation, where the right-hand thumb corresponds to the X-axis, the right-hand index finger corresponds to the Y-axis, and the right-hand middle finger corresponds to the Z-axis. $\mathbf{R}^3$ comes locally equipped with this right-handed orientation at all points. A homeomorphism from $\mathbf{R}^3$ to $\mathbf{R}^3$ is orientation-preserving if the local right-handed frame at each point of the domain maps to a local right-handed frame in the range. Reflection in a hyperplane (such as reflection in the xy-plane by $f:(x,y,z)\rightarrow(x,y,-z)$) reverses the orientation of $\mathbf{R}^3$. We might also require that the circular subspace K come equipped with an orientation (usually indicated by an arrow). If so, we say that our knot or link K is oriented; if not, we say that it is unoriented. Unless otherwise specified, all of our knots will be unoriented. The homeomorphism of pairs h superimposes K_1 on K_2; in this case the knots (links) can be made congruent by a flexible motion or flow (ambient isotopy) of space. An ambient isotopy is a 1-parameter family of homeomorphisms $\{H_t\}_{t=0}^{1}$ of $\mathbf{R}^3$ that begins with the identity and ends with the homeomorphism under consideration: $H_0 = \mathbf{identity}$ and $H_1 = h$. An equivalence class of embeddings is called a knot (link) type.

A knot (link) is usually represented by drawing a diagram (projection) in a plane. This diagram is a shadow of the knot (link) cast on a plane in 3-space. By a small rigid rotation of the knot (link) in 3-space, it can be arranged that no more than two strings cross at any point in the diagram. For short, crossing points in a diagram are called crossings. In the figures in this chapter, at each crossing in a diagram, the undercrossing string is depicted with a break in it, so that the three-dimensional knot (link) type can be uniquely re-created from a two-dimensional diagram. Figure 8.4a-e shows standard diagrams (Rolfsen, 1990) for the knots and links that turn up in Tn3 recombination experiments. In the definition of knot type, we insisted that the transformation that superimposes one knot on another must be orientation-preserving on the ambient space. This restriction allows us to detect a property of great biological significance: chirality. The mirror image of a knot (link) is the configuration obtained by reflecting the configuration in a plane in $\mathbf{R}^3$. Starting with a diagram for a knot (link), one can obtain a diagram for the mirror image by reversing each crossing; the underpass becomes the overpass and vice versa (compare

Figures 8.4d and 8.4e). If K denotes a knot (link), let K^* denote the mirror image. If $K = K^*$, then we say that K is achiral; if $K \neq K^*$, then we say that K is chiral. For example, the Hopf link (Figure 8.4a) and the figure eight knot (Figure 8.4b) are achiral, and the (+) Whitehead

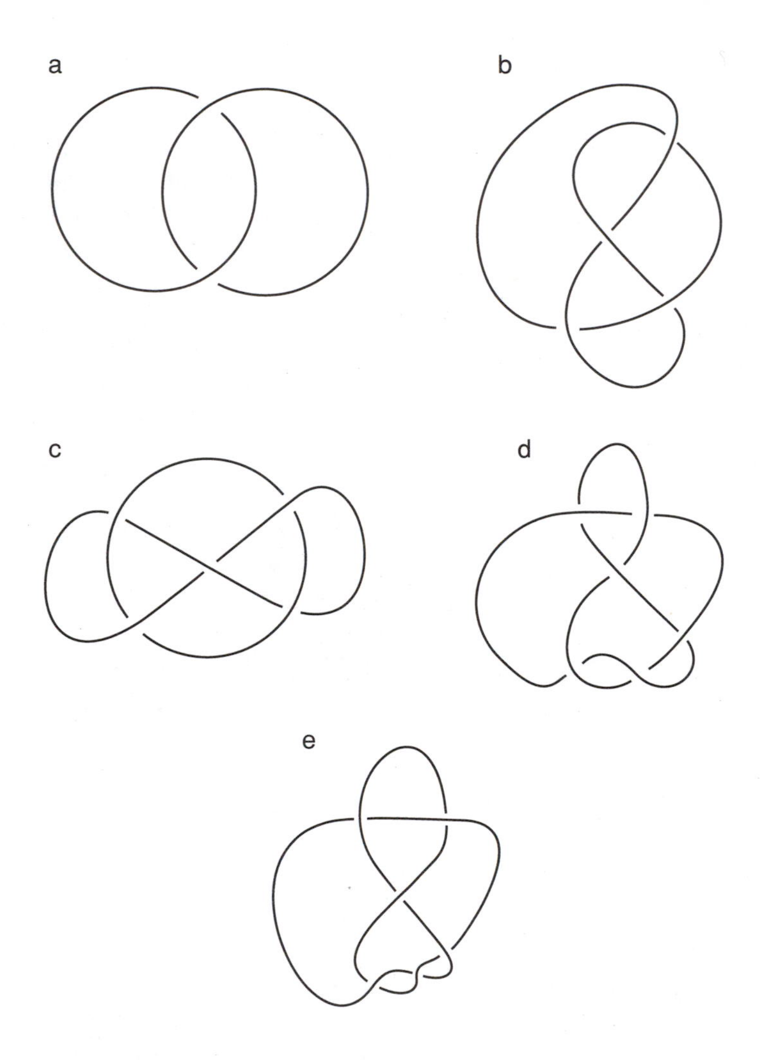

FIGURE 8.4 (a) Hopf link, (b) figure eight knot, (c) (+) Whitehead link, (d) 6_2^*, and (e) 6_2 (mirror image of 6_2^*).

link (Figure 8.4c) and the knot 6_2^* (Figure 8.4d) and its mirror image 6_2 (Figure 8.4e) are chiral. Moreover, all the knots and links in Figure 8.4 are prime, that is, they cannot be formed by the process of tying first one knot in a string and then another.

By moving the knot around in space and then projecting it, it is clear that every given knot (link) type admits infinitely many "different" diagrams, and so the task of recognizing that two completely different diagrams represent the same knot type can be exceedingly difficult. In order to make this job a bit easier, one usually seeks diagrams for the knot type with a minimal number of crossings. This minimal number is called the crossing number of the knot (link) type. The projections in Figures 8.4 are minimal. Crossing number is a topological invariant of knot type. A topological invariant is a number, algebraic group, polynomial, and so on that can be unambiguously attached to a knot (link) type. Most invariants can be algorithmically computed from diagrams (Burde and Zieschang, 1985; Crowell and Fox, 1977; Lickorish, 1988; Kauffman, 1987). If any invariant differs for two knots (links), then the two knots (links) are of different types. If all known invariants are identical, the only conclusion that can be reached is that all known invariants fail to distinguish the candidates. One must then either devise a new invariant that distinguishes the two or prove that they are of the same type by construction of the homeomorphism that transforms one to the other (often by direct geometric manipulation of the diagram or by manipulation of string models). Nevertheless, it is possible to devise invariants (algebraic classification schemes) that uniquely classify certain homologous subfamilies of knots and links, for example, torus knots, two-bridge knots (4-plats), and so on. The algebraic classification schemes for these homologous subfamilies can be used to describe and compute enzyme mechanisms in the topological enzymology protocol.

Fortunately for biological applications, most (if not all) of the circular DNA products produced by in vitro enzymology experiments fall into the mathematically well-understood family of 4-plats. This family consists of knot and link configurations produced by patterns of plectonemic supercoiling of pairs of strands about each other. All "small" knots and links are members of this family—more precisely, all prime knots with crossing number less than 8 and all prime (two-component) links with crossing number less than 7 are 4-plats. A 4-plat is a knot or two-component link that can be formed by platting (or braiding) four strings. All of the knots and links of Figure 8.4 are 4-plats; their standard 4-plat diagrams are shown in Figure 8.5. Each standard 4-plat diagram consists of four horizontal strings, numbered 1 through 4 from top to bottom. The standard pattern of plectonemic interwinding for a 4-plat is encoded by an odd-length classifying vector with positive integer entries $<c_1,c_2,\ldots,c_{2k+1}>$, as shown in Figure 8.5. Beginning from the left, strings in positions 2 and 3 undergo c_1 left-handed plectonemic interwinds (half-twists), then strings in positions 1 and 2 undergo c_2 right-handed plectonemic interwinds, then strings in positions 2 and 3 undergo c_3 left-handed plectonemic interwinds, and this process continues until at the right the strings in positions 2 and 3 undergo c_{2k+1} left-handed plectonemic interwinds. In the standard diagram for a 4-plat, the string in position 4 is not involved in any crossing. The vector representation for the standard diagram of a 4-plat is unique up to reversal of the symbol. That is, the vector $<c_{2k+1},c_{2k},\ldots,c_1>$ represents the same type as the vector $<c_1,c_2,\ldots,c_{2k+1}>$, because turning the 4-plat 180° about the vertical axis reverses the pattern of supercoiling. The standard 4-plat diagram is alternating; that is, as one traverses any strand in the diagram, one alternately encounters over- and undercrossings. Also, standard 4-plat diagrams (with the exception of the unknot $<1>$) are minimal (Ernst and Sumners, 1987).

For in vitro topological enzymology, we can regard the enzyme mechanism as a machine that transforms 4-plats into other 4-plats. We need a mathematical language for describing and computing these enzyme-mediated changes. In many enzyme-DNA reactions, a pair of sites that are distant on the substrate circle are juxtaposed in space and bound to the enzyme. The enzyme then performs its topological moves, and the DNA is then released. We need a mathematical language to describe configurations of linear strings in a spatially confined region. This is accomplished by means of the mathematical concept of tangles. Tangles were introduced into knot theory by J.H. Conway (1970) in a seminal paper involving construction of enumeration schemes for knots and links. The unit 3-ball $\mathbf{B}^3$ in $\mathbf{R}^3$ is the set of all vectors of length ≤ 1. The boundary 2-sphere $\mathbf{S}^2 = \partial\mathbf{B}^3$ is the set of all vectors of length 1. The equator of this 3-ball is the intersection of the boundary $\mathbf{S}^2$ with

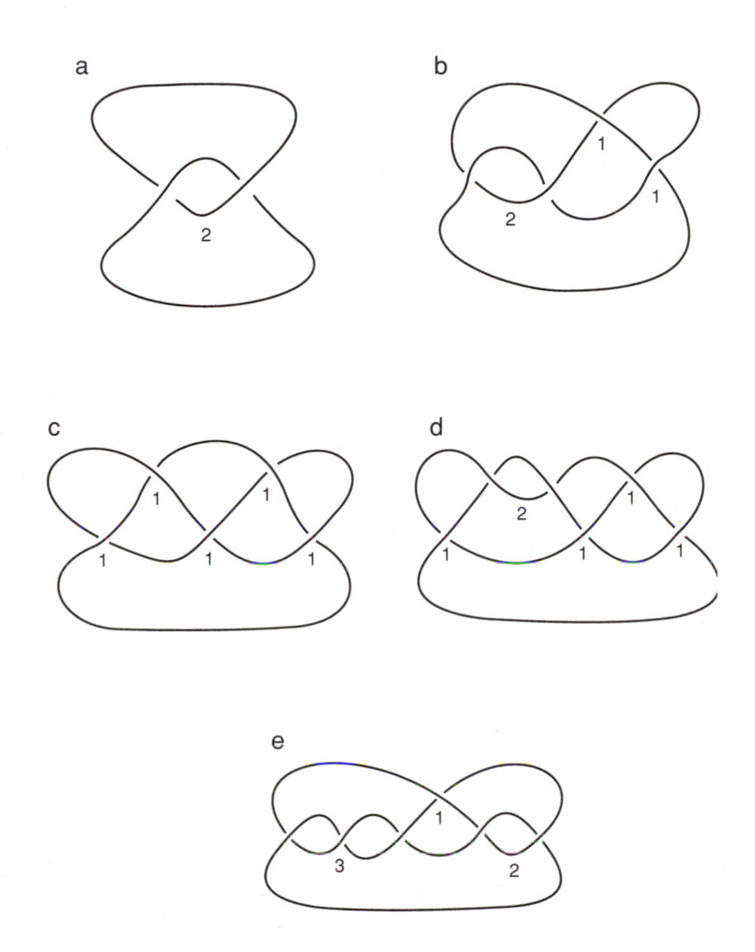

FIGURE 8.5 Standard 4-plats. (a) <2> Hopf link, (b) <2,1,1> figure eight knot, (c) <1,1,1,1,1> (+) figure eight catenane, (d) <1,2,1,1,1> 6_2*, and (e) <3,1,2> 6_2.

the *xy*-plane; the equatorial disk is the intersection of $\mathbf{B}^3$ with the *xy*-plane. On the unit 3-ball, select four points on the equator (called NW, SW, SE, NE). A 2-string tangle in the unit 3-ball is a configuration of two disjoint strings in the unit 3-ball whose endpoints are the four special points {NW,SW,SE,NE}. Two tangles in the unit 3-ball are equivalent if it is possible to elastically transform the strings of one tangle into the strings of the other without moving the endpoints {NW,SW, SE,NE} and without breaking a string or passing one string through another. A class of equivalent tangles is called a tangle type. Tangle theory is knot theory done inside a 3-ball with the ends of the strings firmly glued down. Tangles are usually represented by their projections, called tangle diagrams, onto the equatorial disk in the unit 3-ball, as shown in Figure 8.6. In all figures containing tangles, we assume that the four boundary points {NW,SW,SE,NE} are as in Figure 8.6a, and we suppress these labels.

All four of the tangles in Figure 8.6 are pairwise inequivalent. However, if we relax the restriction that the endpoints of the strings remain fixed and allow the endpoints of the strings to move about on the surface ($\mathbf{S}^2$) of the 3-ball, then the tangle of Figure 8.6a can be trans-

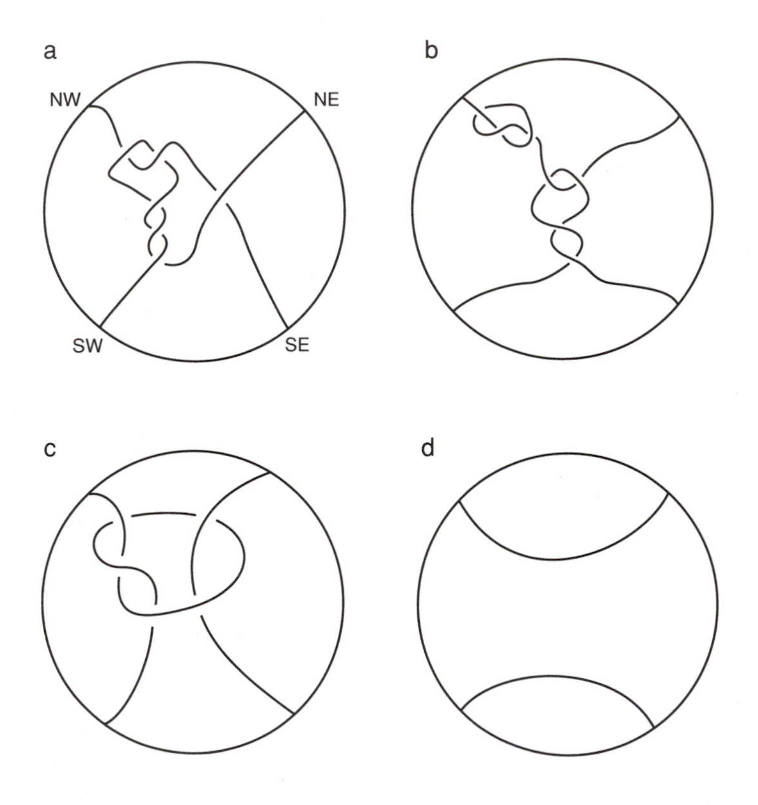

FIGURE 8.6 Tangles. (a) Rational, (b) locally knotted, (c) prime, and (d) trivial.

formed into the trivial tangle of Figure 8.6d. This can be accomplished by rotating (on $\mathbf{S}^2$) the {NE,SE} endpoints one left half-turn (180°) about each other, then rotating the {SW,SE} endpoints three right half-turns about each other, and finally rotating the {NE,SE} endpoints two left half-turns about each other. The tangles in Figures 8.6b and 8.6c cannot be transformed to the trivial tangle by any sequence of such turning motions of the endpoints on $\mathbf{S}^2$. The family of tangles that can be converted to the trivial tangle by moving the endpoints of the strings on $\mathbf{S}^2$ is the family of rational tangles. Equivalently, a rational tangle is one in which the strings can be continuously deformed (leaving the endpoints fixed) entirely into the boundary 2-sphere of the 3-ball, with no string passing through itself or through another string.

Rational tangles form a homologous family of 2-string configurations in $\mathbf{B}^3$ and are formed by a pattern of plectonemic supercoiling of pairs of strings. Like 4-plats, rational tangles look like DNA configurations, being built up out of plectonemic supercoiling of pairs of strings. More specifically, enzymes are often globular in shape and are topologically equivalent to our unit defining ball $\mathbf{B}^3$. Thus, in an enzymatic reaction between a pair of DNA duplexes, the pair {enzyme, bound DNA} forms a 2-string tangle. Since the amount of bound DNA is small, the enzyme-DNA tangle so formed will admit projections with few nodes and therefore is very likely rational. For example, all locally unknotted 2-string tangles having less than five crossings are rational. There is a second, more natural argument for rationality of the enzyme-DNA tangle. In all cases studied intensively, DNA is bound to the surface of the protein. This means that the resulting protein-DNA tangle is rational, since any tangle whose strings can be continuously deformed into the boundary of the defining ball is automatically rational.

A classification scheme for rational tangles is based on a standard form that is a minimal alternating diagram. The classifying vector for a rational tangle is an integer-entry vector $(a_1, a_2, \ldots, a_n)$ of odd or even length, with all entries (except possibly the last) nonzero and having the same sign, and with $|a_1| > 1$. The integers in the classifying vector represent the left-to-right (west-to-east) alternation of vertical and horizontal windings in the standard tangle diagram, always ending with horizontal windings on the east side of the diagram. Horizontal winding is the winding between strings in the top and bottom (north and south) positions; vertical winding is the winding between strings in the left and right (west and east) positions. By convention, positive integers correspond to horizontal plectonemic right-handed supercoils and vertical left-handed plectonemic supercoils; negative integers correspond to horizontal left-handed plectonemic supercoils and vertical right-handed plectonemic supercoils. Figure 8.7 shows some standard tangle diagrams. Two rational tangles are of the same type if and only if they have identical classifying vectors. Due to the requirement that $|a_1| > 1$ in the classifying vector convention for rational tangles, the corresponding tangle projection must have at least two nodes. There are four rational tangles $\{(0),(0,0),(1),(-1)\}$ that are exceptions to this convention ($|a_1| = 0$ or 1) and are displayed in Figure 8.7c-f. The classifying vector $(a_1, a_2, \ldots, a_n)$ can be converted to an (extended) rational number $b/a \in Q \cup \infty$ by means of the following continued fraction calculation:

$$b/a \;=\; a_n + 1/(a_{n-1} + (1/(a_{n-2} + \cdots))) \,.$$

Two rational tangles are of the same type if and only if these (extended) rational numbers are equal (Conway, 1970), which is the reason for calling them "rational" tangles.

In order to use tangles as building blocks for knots and links, and mathematically to mimic enzyme action on DNA, we now introduce the geometric operations of tangle addition and tangle closure. Given tangles *A* and *B*, one can form the tangle $A + B$ as shown in Figure 8.8a. The sum of two rational tangles need not be rational. Given any tangle *C*, one can form the closure *N*(*C*) as in Figure 8.8b. In the closure operation on a 2-string tangle, ends NW and NE are connected, ends SW and SE are connected, and the defining ball is deleted, leaving a knot or a link of two components. Deletion of the defining $\mathbf{B}^3$ is analogous to deproteinization of the DNA when the synaptosome dissociates. One can combine the operations of tangle addition and tangle closure to create a tangle equation of the form *N*(*A* + *B*) = knot (link). In such a tangle equation, the tangles *A* and *B* are said to be summands of the resulting knot (link). An example of this phenomenon is the tangle equation $N((-3,0) + (1)) = <2>$, shown in Figure 8.8c. In general, if *A* and *B* are any two rational tangles, then

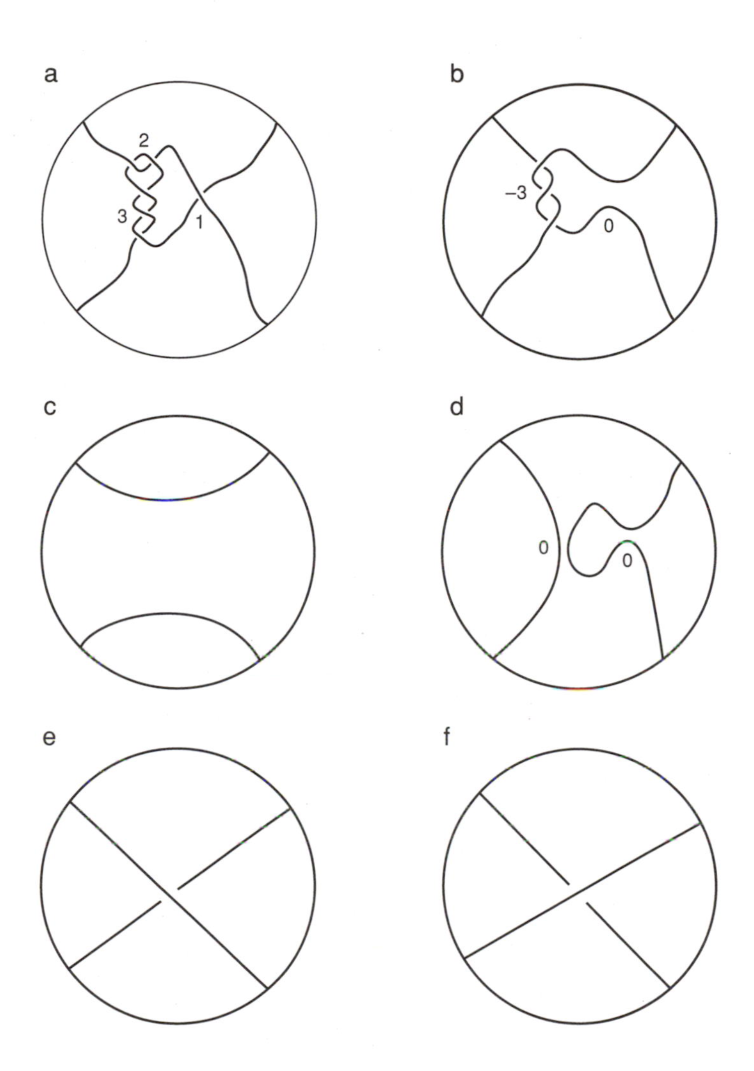

FIGURE 8.7 Tangle diagrams. (a) (2,3,1), (b) (−3,0), (c) (0), (d) (0,0), (e) (1), and (f) (−1).

$N(A+B)$ is a 4-plat. Given these constructions, rational tangles are summands for 4-plats.

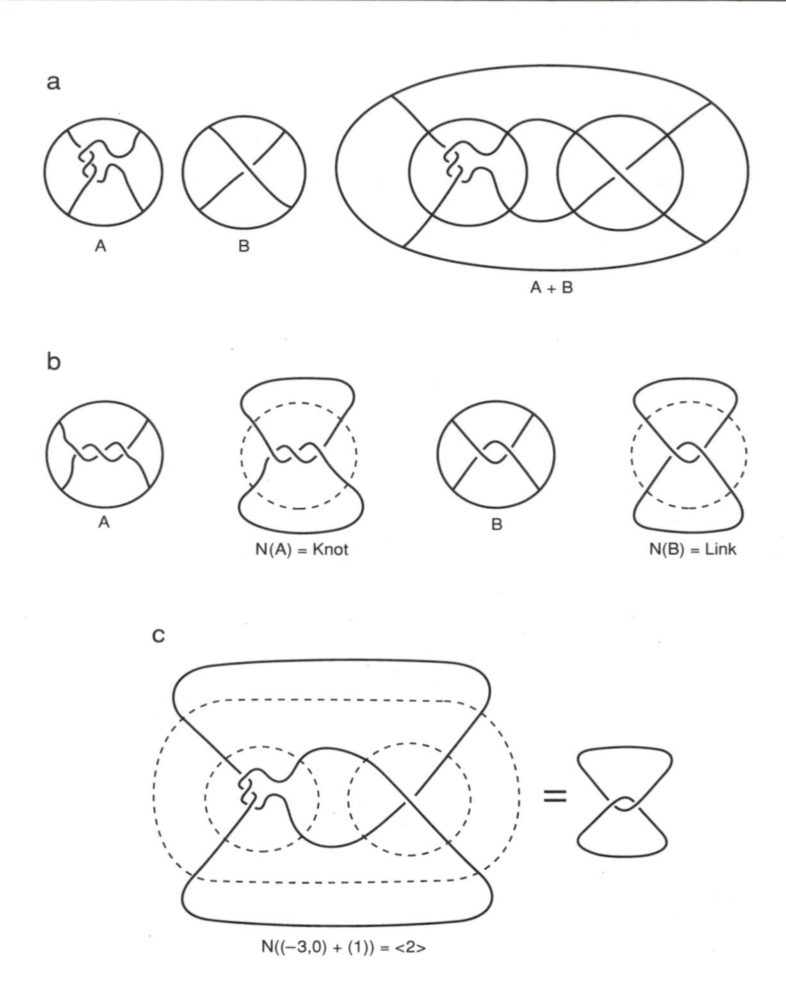

FIGURE 8.8. Tangle operations. (a) Tangle addition, (b) tangle closure, and (c) $N((-3,0)+(1)) =< 2 >$

THE TANGLE MODEL FOR SITE-SPECIFIC RECOMBINATION

The fundamental observations underlying this model are that a pair of sites bound by an enzyme forms a tangle and that most of the products of recombination experiments performed on unknotted substrate are 4-

plats. We will use tangles to build a model that will compute the topology of the pre- and post-recombination synaptic complex in a single recombination event, given knowledge of the topology of the substrate and product (Ernst and Sumners, 1990; Sumners, 1990, 1992; Sumners et al., 1994). In site-specific recombination on circular DNA substrate, two kinds of geometric manipulation of the DNA occur. The first is a global ambient isotopy, in which a pair of distant recombination sites are juxtaposed in space, and the enzyme binds to the molecule(s), forming the synaptic complex. Once synapsis is achieved, the next move is local and due entirely to enzyme action. Within the region occupied by the enzyme, the substrate is broken at each site, and the ends are recombined. We will model this local move.

The aim of our mathematical model is, given the observed changes in geometry and topology of the DNA, to compute the topology of the entire synaptic complex, both before and after enzyme action. Within the region controlled by the enzyme, the enzyme breaks the DNA at each site and recombines the ends by exchanging them. We model the enzyme itself as a 3-ball. The synaptosome consisting of the enzyme and bound DNA forms a 2-string tangle.

What follows is a list of biological and mathematical assumptions made in the tangle model (Ernst and Sumners, 1990; Sumners, 1992; Sumners et al., 1994). Most of these assumptions are implicit in the existing analyses of the results of enzyme experiments on circular DNA (Cozzarelli et al., 1984; Stark et al., 1989; Spengler et al., 1985; Wasserman and Cozzarelli, 1986; Wasserman et al., 1985; Kanaar et al., 1990; White et al., 1987; Kanaar et al., 1988; Abremski et al., 1986; Droge and Cozzarelli, 1986; Spengler et al., 1984).

We make the following biological assumption:

Assumption 1 *The enzyme mechanism in a single recombination event is constant, independent of the geometry (supercoiling) and topology (knotting and catenation) of the substrate population. Moreover, recombination takes place entirely within the domain of the enzyme ball, and the substrate configuration outside the enzyme ball remains fixed while the strands are being broken and recombined inside and on the boundary of the enzyme.*

That is, we assume that any two pre-recombination copies of the synaptosome are identical, meaning that we can by rotation and translation superimpose one copy on the other, with the congruence so achieved respecting the structure of both the protein and the DNA. We likewise assume that all of the copies of the post-recombination synaptosome are identical.

In a recombination event, we can mathematically divide the DNA involved into three types: (1) the DNA at and very near the sites where the DNA breakage and reunion are taking place; (2) other DNA bound to the enzyme, which is unchanged during a recombination event; and (3) the DNA in the synaptic complex that is not bound to the enzyme and that does not change during recombination. We make the following mathematical assumption about DNA types (1) and (2):

Assumption 2 *The synaptosome is a 2-string tangle and can be mathematically subdivided into the sum* $O_b + P$ *of two tangles.*

One tangle, the parental tangle *P*, contains the recombination sites where strand breakage and reunion take place. The other tangle, the outside bound tangle O_b, is the remaining DNA in the synaptosome outside the *P* tangle; this is the DNA that is bound to the enzyme but that remains unchanged during recombination. The enzyme mechanism is modeled as tangle replacement (surgery) in which the parental tangle *P* is removed from the synaptosome and replaced by the recombinant tangle *R*. Therefore, our model assumes the following:

$$\begin{aligned}\text{pre-recombination synaptosome} &= O_b + P \\ \text{post-recombination synaptosome} &= O_b + R\,.\end{aligned}$$

In order to accommodate nontrivial topology in the DNA of type (3), we let the outside free tangle O_f denote the synaptic complex DNA that is free (not bound to the enzyme) and that is unchanged during a single recombination event. We make the following mathematical assumption:

Assumption 3 *The entire synaptic complex is obtained from the tangle sum* (O_f *+ synaptosome) by the tangle closure construction.*

If one deproteinizes the pre-recombination synaptic complex, one obtains the substrate; deproteinization of the post-recombination synaptic complex yields the product. The topological structure (knot and catenane types) of the substrate and product yields equations in the recombination variables $\{O_f, O_b, P, R\}$. Specifically, a single recombination event on a single circular substrate molecule produces two recombination equations in four unknowns:

$$\text{substrate equation:} \quad N(O_f + O_b + P) = \text{substrate}$$
$$\text{product equation:} \quad N(O_f + O_b + R) = \text{product}.$$

The geometric meaning of these recombination equations is illustrated in Figure 8.3. In Figure 8.3, $O_f = (0)$, $O_b = (-3,0)$, $P = (0)$, and $R = (1)$. With these values for the variables, our recombination equations become:

$$\text{substrate equation:} \quad N((0) + (-3,0) + (0)) = <1>$$
$$\text{product equation:} \quad N((0) + (-3,0) + (1)) = <2>.$$

THE TOPOLOGY OF Tn3 RESOLVASE

Tn3 resolvase is a site-specific recombinase that reacts with certain circular duplex DNA substrate with directly repeated recombination sites (Wasserman et al., 1985). One begins with supercoiled unknotted DNA substrate and treats it with resolvase. The principal product of this reaction is known to be the DNA 4-plat $<2>$ (the Hopf link, Figures 8.4a and 8.5a) (Wasserman and Cozzarelli, 1985). Resolvase is known to act dispersively in this situation—to bind to the circular DNA, to mediate a single recombination event, and then to release the linked product. It is also known that resolvase and free (unbound) DNA links do not react. However, once in 20 encounters, resolvase acts processively—additional recombinant strand exchanges are promoted prior to the release of the product, with yield decreasing exponentially with increasing number of strand exchanges at a single binding encounter with the enzyme. Two successive rounds of processive recombination produce the DNA 4-plat

$<2,1,1>$ (the figure eight knot, Figures 8.4b and 8.5b); three successive rounds of processive recombination produce the DNA 4-plat $<1,1,1,1,1>$ (the Whitehead link, Figures 8.4c and 8.5c), whose electron micrograph appears in Figure 8.1b; four successive rounds of recombination produce the DNA 4-plat $<1,2,1,1,1>$ (the knot 6_2^*, Figures 8.4d and 8.5d), whose electron micrograph appears in Figure 8.1c. The discovery of the DNA knot $<1,2,1,1,1>$ substantiated a model for Tn3 resolvase mechanism (Wasserman et al., 1985).

In processive recombination, it is the synaptosome itself that repeatedly changes structure. We make the following biologically reasonable mathematical assumption in our model:

Assumption 4 *In processive recombination, each additional round of recombination adds a copy of the recombinant tangle R to the synaptosome.*

More precisely, n rounds of processive recombination at a single binding encounter generate the following system of $(n+1)$ tangle equations in the unknowns $\{O_f, O_b, P, R\}$:

$$\begin{aligned} &\text{substrate:} && N(O_f + O_b + P) = \text{substrate} \\ &r\text{th round:} && N(O_f + O_b + rR) = r\text{th round product, } 1 \le r \le n. \end{aligned}$$

For resolvase, the electron micrograph of the synaptic complex in Figure 8.2 reveals that $O_f = (0)$, since the DNA loops on the exterior of the synaptosome can be untwisted and are not entangled. This observation from the micrograph reduces the number of variables in the tangle model by one, leaving us with three variables $\{O_b, P, R\}$. One can prove (Sumners, 1990, 1992; Ernst and Sumners, 1990) that there are four possible tangle pairs $\{O_b, R\}$, which can produce the experimental results of the first two rounds of processive Tn3 recombination. The third round of processive recombination is then used to discard three of these four pairs as extraneous solutions. The following theorems can be viewed as a mathematical proof of resolvase synaptic complex structure: the model proposed in (Wasserman et al., 1985) is the unique explanation for the first three observed products of processive Tn3 recombination, assuming that processive recombination acts by adding on copies of the recombinant tangle R.

The process of obtaining electron micrographs of RecA-enhanced DNA knots and catenanes is technically difficult and requires a relatively large amount of product due to the extensive work-up required for RecA coating and microscopy. Gel electrophoresis is not only technically much easier to do, but it detects vanishingly small amounts of DNA product. For these reasons, biologists prefer to use gel electrophoresis as the assay from which experimental conclusions are to be drawn. For relaxed DNA knots and links, the gel determines the crossing number of the (relaxed) products, and comparison to gel ladders for known knot and catenane structures can be used to obtain more information than crossing number alone. As an aid to the analysis of topological enzymology experiments, a table of possible (and biologically reasonable!) tangle mechanisms has been prepared (Sumners et al., 1994) for each possible sequence of crossing numbers of reaction products that can be read from the gel. This tangle table should make the mathematical analysis of topological enzymology experiments easier to do.

We now come to the rigorous mathematical proof of Tn3 mechanism. The proofs of the following two theorems can be skipped without detriment to the continuity of the exposition.

Theorem 8.1 *Suppose that tangles* O_b, P, *and* R *satisfy the following equations:*

(i) $N(O_b + P) = <1>$ *(the unknot)*

(ii) $N(O_b + R) = <2>$ *(the Hopf link)*

(iii) $N(O_b + R + R) = <2,1,1>$ *(the figure 8 knot).*

Then $\{O_b, R\} = \{(-3,0),(1)\}$, $\{(3,0),(-1)\}$, $\{(-2,-3,-1),(1)\}$, *or* $\{(2,3,1),(-1)\}$.

Proof: In this proof we use the following notation: $\mathbf{R}^n$ denotes Euclidean n-space, $\mathbf{B}^n$ denotes the unit ball in $\mathbf{R}^n$ (the set of all vectors in $\mathbf{R}^n$ of length ≤ 1), and $\mathbf{S}^{n-1}$ denotes the boundary of $\mathbf{B}^n$ (the set of all vectors in $\mathbf{R}^n$ of length 1). The first (and mathematically most interesting) step in the proof of this theorem is to argue that solutions $\{O_b, R\}$ must be rational tangles. Now O_b, R, and $(O_b + R)$ are locally unknotted, because $N(O_b + R)$ is the Hopf link, which has two unknotted components. Any local knot in a tangle summand would persist in the Hopf link. Likewise, P is locally unknotted, because $N(O_b + P)$ is the unknot. Let A' denote the 2-fold branched cyclic cover of the tangle A; then $\partial A' = \mathbf{S}^1 \times \mathbf{S}^1$. If A is a prime tangle, then the inclusion homeomorphism injects $\pi_1(\partial A') = \mathbf{Z} \oplus \mathbf{Z}$ into $\pi_1(A')$ (Lickorish, 1981). If both A and B are prime tangles, and $N(A+B)'$ denotes the 2-fold branched cyclic cover, then $\pi_1(N(A+B)')$ contains a subgroup isomorphic to $\mathbf{Z} \oplus \mathbf{Z}$. If K is any 4-plat, then $\pi_1(K')$ is a cyclic group, since K' is a lens space (Burde and Zieschang, 1985). Since no cyclic group contains $\mathbf{Z} \oplus \mathbf{Z}$, no 4-plat has two prime tangle summands. This means that if A and B are locally unknotted tangles, and $N(A+B)$ is a 4-plat, then *at least one* of A and B must be a rational tangle. From equation (ii) above, we conclude that at least one of $\{O_b, R\}$ is rational. Suppose that O_b is rational and that R is prime. Given that $N((O_b + R) + R)$ is a knot, one can argue (Lickorish, 1981) that $O_b + R$ is also a prime tangle. From equation (iii), we then have that the 4-plat $< 2,1,1 >$ admits two prime tangle summands, which is impossible. We therefore conclude that R must be a rational tangle.

The next step is to argue that O_b is a rational tangle. Suppose that O_b is a prime tangle. Then P must be a rational tangle, because $N(O_b + P)$ is the unknot (equation (i)). Passing to 2-fold branched cyclic covers, we have that $N(O_b + P)' = \mathbf{S}^3$, and P' is homeomorphic to $\mathbf{S}^1 \times \mathbf{B}^2$ (since P is rational), so O_b' is a bounded knot complement in $\mathbf{S}^3$. We know that R is a rational tangle and can argue that equation (iii) implies that $(R+R)$ is likewise rational. Again passing to the 2-fold branched cyclic covers of equations (ii) and (iii), we obtain the equations $N(O_b + R)' =$ the lens space $L(2,1)$ and $N(O_b + (R+R))' =$ the lens space $L(5,3)$. Since R' and $(R+R)'$ are each homeomorphic to a solid torus $\mathbf{S}^1 \times \mathbf{B}^2$, this means that there are two attachments of a solid torus to O_b' along $\partial O_b' = \mathbf{S}^1 \times \mathbf{S}^1$, yielding the lens spaces $L(2,1)$ and $L(5,3)$. The process of adding on a solid torus along its boundary is called Dehn surgery, and the Cyclic Surgery Theorem (Culler et al., 1987) now applies to this situation to imply that, since the orders of the cyclic fundamental groups of the lens spaces differ by more than one, the only way this can happen is for O_b' to be a Seifert fiber space and hence a torus knot complement. Fortunately, the results of Dehn surgery on torus knot complements are well understood, and one can argue that in fact O_b' must be a complement of the unknot (a solid torus) (Ernst and Sumners, 1990), which means that O_b is a rational tangle.

The proof now amounts to computing the rational solutions to equations (ii) and (iii), exploiting the classifying schemes for rational tangles and 4-plats. In Ernst and Sumners (1990), a "calculus for rational tangles" was developed to perform such calculations. One can use this calculus of classifying vectors to solve equations (ii) and (iii), obtaining the four solution pairs $\{O_b, R\} = \{(-3,0),(1)\}$, $\{(3,0),(-1)\}$, $\{(-2,-3,-1),(1)\}$, and $\{(2,3,1),(-1)\}$. Because each of the unoriented 4-plat products in equations (ii) and (iii) is achiral, given any solution set $\{O_b, R\}$ to equations (ii) and (iii), its mirror image $\{-O_b, -R\}$ must also be a solution. So the mathematical situation, given equations (i) through (iii), is that we have two pairs of mirror image solution sets for $\{O_b, R\}$.

In order to decide which is the biologically correct solution, we must utilize more experimental evidence. The third round of processive resolvase recombination determines which of these four solutions is the correct one.

Theorem 8.2 *Suppose that tangles O_b, P, and R satisfy the following equations:*

(i) $N(O_b + P) = <1>$ *(the unknot)*

(ii) $N(O_b + R) = <2>$ *(the Hopf link)*

(iii) $N(O_b + R + R) = <2,1,1>$ *(the figure 8 knot).*

(iv) $N(O_b + R + R + R) = <1,1,1,1,1>$ *(the (+) Whitehead link).*

Then

$O_b = -3,0,\ R = (1),$ *and* $N(O_b + R + R + R + R) = <1,2,1,1,1>$.

Proof: The unoriented (+) Whitehead link is chiral and $\{O_b R\} = \{(-3,0),(1)\}$ is the unique solution to equations (i) and (iv).

The correct global topology of the first round of processive Tn3 recombination on the unknot is shown in Figure 8.3. Moreover, the first three rounds of processive Tn3 recombination uniquely determine $N(O_b + R + R + R + R)$, the result of four rounds of recombination. It is the 4-plat knot <1,2,1,1,1>, and this DNA knot has been observed (Figure 8.1c). We note that there is no information in either Theorem 8.1 or Theorem 8.2 about the parental tangle *P*. Since *P* appears in only one tangle equation (equation (i)), for each fixed rational tangle solution for O_b, there are infinitely many rational tangle solutions to equation (i) for *P* (Ernst and Sumners, 1990). Most biologists believe that $P = (0)$, and a biomathematical argument exists for this claim (Sumners et al., 1994).

SOME UNSOLVED PROBLEMS

1. *How does TOPO II recognize knots? E. coli* contains circular duplex DNA molecules. In wild-type *E. coli*, no knotting has been observed for these molecules. However, in a mutant strain of *E. coli* where the production of Topoisomerase II (the enzyme that performs strand passage via an enzyme-bridged transient double-stranded break in the DNA) can be blocked by heat shock, a small fraction (about 7 percent) of knotted DNA has been observed (Shishido et al., 1987). All observed knots have the gel mobility of trefoil knots. The observed knots are presumably the by-products of other cellular processes (such as recombination). This experiment shows that TOPO II is able to detect DNA knots and kill them in wild-type *E. coli*. How does the enzyme (which can act only locally) detect the global topology of a DNA knot and then make just the right combination of passages to kill the knot? It must be the energy minimization of the DNA itself that detects the knotting. The enzyme has only to detect when two DNA strands are being pushed together in space by the DNA itself in an effort to attain a lower energy state, whence the enzyme can operate, allowing one DNA strand to pass through another to reach a lower-energy configuration. If one ties a knot in a short stiff rubber tube (an elastic tube) and seals up the ends to form a knotted circle, the tube will touch itself, trying to pass through itself to

relieve strain and minimize energy. For circular elastica in $\mathbf{R}^3$, minimization of the bending energy functional occurs when the elasticum is the round planar unknot (Langer and Singer, 1984, 1985). This means that a knotted elasticum has at least one point of self-contact and that the elasticum is pushing at that point of self-contact to get through to a lower energy state. Does this elasticum model adequately explain the ability of Topoisomerase II to detect and selectively kill DNA knots in vivo?

2. *What is the topology of the kDNA network?* The kinetoplast DNA (kDNA) of the parasite *trypanosome* forms a link of some 5,000 to 10,000 unknotted DNA circles—the DNA equivalent of chain mail (Marini et al., 1980; Englund et al., 1982; Rauch et al., 1994). Work is ongoing (Rauch et al., 1994) in which the topological structure of kDNA is being studied by means of partial digest of the network, electrophoresis, and electron microscopy of the characteristic fragments, in which the large kDNA link is being randomly broken up into small sublinks, and the frequency of occurrence of these sublink units is being used (statistically) to reconstruct the large link itself. The kDNA network consists of small minicircles and a few large maxicircles. The minicircles are known to be unknotted, and it is known that neighbors link in the fashion of the Hopf link (Figure 8.4a) (like the links in a chain). Moreover, it is believed that the kDNA network has a fundamental region that is repeated in space to generate the entire structure. This gives rise to a knot theory problem: classify the links that allow a diagram in which each component has no self-crossings (and hence is unknotted) and in which each component links another component simply (like the links in a chain) or not at all, and in which the linking structure is periodic in space. The spatial periodicity amounts to drawing the link diagram on a torus (or some other compact, orientable 2-manifold), from whence the entire diagram is reproduced by taking the universal cover. The algebraic classification of such "chain mail links" should be interesting and obtainable with off-the-shelf topological invariants. Another topological problem has arisen in this biological system. A trefoil knotted minicircle has been observed as an intermediate to the replication process on the kDNA network (Ryan et al., 1988). What is the mechanism that produces this knotted minicircle? Does the topology of the network naturally generate knots as replication intermediates?

3. *Why is the figure eight knot faster than the trefoil knot?* The phenomenon of gel mobility of relaxed knotted duplex DNA circles (Dean et al., 1985) has no adequate theoretical explanation. The gel velocity of

relaxed DNA knots is determined by crossing number; the larger the crossing number, the faster the migration. Perhaps this is because among knots of the same length with small crossing numbers, the average value of the radius of gyration (a measure of the average size) correlates strongly with crossing number. It is very curious that the crossing number, clearly an artifact of planar diagrammatic representation of knots, would have anything at all to do with the three-dimensional average knot confor-mation. What is the relationship (if any) between radius of gyration of DNA circles of fixed molecular weight and fixed knot type, crossing number, and the gel mobility of these knotted DNA circles?

ANNOTATED BIBLIOGRAPHY

Knot Theory

Adams, C., 1994, *The Knot Book: An Elementary Introduction to Mathematical Theory of Knots,* New York: W.H. Freeman.

Kauffman, L.H., 1987, *On Knots,* Princeton, N.J.: Princeton University Press.

Livingston, C., 1994, *Knot Theory, Carus Mathematical Monograph,* Vol. 24, Washington, D.C.: Mathematical Association of America.

Rolfsen, D., 1990, *Knots and Links,* Berkeley, Calif.: Publish or Perish, Inc.

Each of these mathematics books has an easygoing, reader-friendly style and numerous pictures, a very important commodity when one is trying to understand knot theory.

Application of Geometry and Topology to Biology

Bauer, W.R., F.H.C. Crick, and J.H. White, 1980, "Supercoiled DNA," *Scientific American* **243**, 100-113.

This paper is a very nice introduction to the description and measurement of DNA supercoiling.

Sumners, D.W., 1987, "The role of knot theory in DNA research," pp. 297-318 in *Geometry and Topology,* C. McCrory and T. Shifrin (eds.), New York: Marcel Dekker.

Sumners, D.W., 1990, "Untangling DNA," *The Mathematical Intelligencer* **12**, 71-80.

These papers are expository articles written for a mathematical audience. The first gives an overview of knot theory and DNA, and the second describes the tangle model.

Sumners, D.W. (ed.), 1994, *New Scientific Applications of Geometry and Topology, Proceedings of Symposia in Applied Mathematics,* Vol. 45, Providence, R.I.: American Mathematical Society.

This volume contains six expository papers outlining new applications of geometry and topology in molecular biology, chemistry, polymers, and physics. Three of the papers concern DNA applications.

Walba, D.M., 1985, "Topological stereochemistry," *Tetrahedron* **41**, 3161-3212.

This paper is written by a chemist and describes topological ideas in synthetic chemistry and molecular biology. It is a good place to witness the translation of technical terms of science to mathematical concepts, and vice versa.

Wang, J.C., 1982, "DNA topoisomerases," *Scientific American* **247**, 94-109.

This paper describes how topoisomerases act to control DNA geometry and topology in various life processes in the cell.

Wasserman, S.A., and N.R. Cozzarelli, 1986, "Biochemical topology: Applications to DNA recombination and replication," *Science* **232**, 951-960.

This paper describes the topological approach to enzymology protocol and reviews the results of various experiments on topoisomerases and recombinases.

White, J.H., 1989, "An introduction to the geometry and topology of DNA structure," pp. 225-253 in *Mathematical Methods for DNA Sequences,* M.S. Waterman (ed.), Boca Raton, Fla.: CRC Press.

This is a very nice introductory mathematical treatment of linking number, twist, and writhe, with DNA applications.

REFERENCES

Abremski, K., B. Frommer, and R.H. Hoess, 1986, "Linking-number changes in the DNA substrate during Cre-mediated loxP site-specific recombination," *Journal of Molecular Biology* **192**, 17-26.

Benjamin, H.W., and N.R. Cozzarelli, 1990, "Geometric arrangements of Tn3 resolvase sites," *J. Biol. Chem.* **265**, 6441-6447.

Burde, G., and H. Zieschang, 1985, *Knots,* New York: W. De Gruyter.

Conway, J.H., 1970, "An enumeration of knots and links and some of their related properties," pp. 329-358 in *Computational Problems in Abstract Algebra; Proceedings of a Conference at Oxford 1967,* Oxford: Pergamon Press.

Cozzarelli, N.R., 1992, "The biological roles of DNA topology," in *New Scientific Applications of Geometry and Topology, Proceedings of Symposia in Applied Mathematics,* D.W. Sumners (ed.), Providence, R.I.: American Mathematical Society.

Cozzarelli, N.R., M.A. Krasnow, S.P. Gerrard, and J.H. White, 1984, "A topological treatment of recombination and topoisomerases," *Cold Spring Harbor Symp. Quant. Biol.* **49**, 383-400.

Crowell, R.H., and R.H. Fox, 1977, *Introduction to Knot Theory. Graduate Texts in Mathematics* **57**, New York: Springer-Verlag.

Culler, M.C., C.M. Gordon, J. Luecke, and P.B. Shalen, 1987, "Dehn surgery on knots," *Ann. Math.* **125**, 237-300.

Dean, F.B., A. Stasiak, T. Koller, and N.R. Cozzarelli, 1985, "Duplex DNA knots produced by *Escherichia coli* topoisomerase I," *J. Biol. Chem.* **260**, 4975-4983.

Droge, P., and N.R. Cozzarelli, 1989, "Recombination of knotted substrates by Tn3 resolvase," *Proceedings of the National Academy of Sciences USA* **86**, 6062-6066.

Englund, P.T., S.L. Hajduk, and J.C. Marini, 1982, "The molecular biology of trypanosomes," *Annu. Rev. Biochem.* **51**, 695-726.

Ernst, C., and D.W. Sumners, 1987, "The growth of the number of prime knots," *Math. Proc. Cambridge Philos. Soc.* **102**, 303-315.

Ernst, C., and D.W. Sumners, 1990, "A calculus for rational tangles: Applications to DNA recombination," *Math. Proc. Cambridge Philos. Soc.* **108**, 489-515.

Griffith, J.D., and H.A. Nash, 1985, "Genetic rearrangement of DNA induces knots with a unique topology: Implications for the mechanism of synapsis and crossing-over," *Proceedings of the National Academy of Sciences USA* **82**, 3124-3128.

Heichman, K.A., and R.C. Johnson, 1990, "The Hin invertasome: Protein-mediated joining of distant recombination sites at the enhancer," *Science* **249**, 511-517.

Kanaar, R., P. van de Putte, and N.R. Cozzarelli, 1988, "Gin-mediated DNA inversion: Product structure and the mechanism of strand exchange," *Proceedings of the National Academy of Sciences USA* **85**, 752-756.

Kanaar, R., A. Klippel, E. Shekhtman, J.M. Dungan, R. Kahmann, and N.R. Cozzarelli, 1990, "Processive recombination by the phage Mu gin system: Implications for mechanisms of DNA exchange, DNA site alignment, and enhancer action," *Cell* **62**, 353-366.

Kauffman, L.H., 1987, *On Knots,* Princeton, N.J.: Princeton University Press.

Kim, S., and A. Landy, 1992, "Lambda Int protein bridges between higher order complexes at two distant chromosomal loci *att*L and *att*R," *Science* **256**, 198-203.

Krasnow, M.A., A. Stasiak, S.J. Spengler, F. Dean, T. Koller, and N.R. Cozzarelli, 1983, "Determination of the absolute handedness of knots and catenanes of DNA," *Nature* **304**, 559-560.

Langer, J., and D.A. Singer, 1984, "Knotted elastic curves in R^3," *J. London Math. Soc.* **30**, 512-520.

Langer, J., and D.A. Singer, 1985, "Curve straightening and a minimax argument for closed elastic curves," *Topology* **24**, 75-88.

Lickorish, W.B.R., 1981, "Prime knots and tangles," *Trans. Am. Math. Soc.* **267**, 321-332.

Lickorish, W.B.R., 1988, "Polynomials for links," *Bull. London Math. Soc.* **20**, 558-588.

Marini, J.C., K.G. Miller, and P.T. Englund, 1980, "Decatenation of kinetoplast DNA by topoisomerases," *J. Biol. Chem.* **255**, 4976-4979.

Pollock, T.J., and H.A. Nash, 1983, "Knotting of DNA caused by genetic rearrangement: Evidence for a nucleosome-like structure in site-specific recombination of bacteriophage lambda," *Journal of Molecular Biology* **170**, 1-18.

Rauch, C.A., P.T. Englund, S.J. Spengler, N.R. Cozzarelli, and J.H. White, 1994, "Kinetoplast DNA: Structure and replication," (in preparation).

Rolfsen, D., 1990, *Knots and Links,* Berkeley, Calif.: Publish or Perish, Inc.

Ryan, K.A., T.A. Shapiro, C.A. Rauch, J.D. Griffith, and P.T. Englund, 1988, "A knotted free minicircle in kinetoplast DNA," *Proceedings of the National Academy of Sciences USA* **85**, 5844-5848.

Sherratt, D., P. Dyson, M. Boocock, L. Brown, D. Summers, G. Stewart, and P. Chan, 1984, "Site-specific recombination in transposition and plasmid stability," *Cold Spring Harbor Symp. Quant. Biol.* **49**, 227-233.

Shishido, K., N. Komiyama, and S. Ikawa, 1987, "Increased production of a knotted form of plasmid pBR322 DNA in *Escherichia coli* DNA topoisomerase mutants," *Journal of Molecular Biology* **195**, 215-218.

Spengler, S.J., A. Stasiak, and N.R. Cozzarelli, 1984, "Quantitative analysis of the contributions of enzyme and DNA to the structure of lambda integrative recombinants," *Cold Spring Harbor Symp. Quant. Biol.* **49**, 745-749.

Spengler, S.J., A. Stasiak, and N.R. Cozzarelli, 1985, "The stereostructure of knots and catenanes produced by phage lambda integrative recombination: Implications for mechanism and DNA structure," *Cell* **42**, 325-334.

Stark, W.M., D.J. Sherratt, and M.R. Boocock, 1989, "Site-specific recombination by Tn3 resolvase: Topological changes in the forward and reverse reactions," *Cell* **58**, 779-790.

Sumners, D.W., 1987a, "Knots, macromolecules and chemical dynamics," pp. 297-318 in *Graph Theory and Topology in Chemistry,* King and Rouvray (eds.), New York: Elsevier.

Sumners, D.W., 1987b, "The role of knot theory in DNA research," pp. 297-318 in *Geometry and Topology,* C. McCrory and T. Shifrin (eds.), New York: Marcel Dekker.

Sumners, D.W., 1990, "Untangling DNA," *The Mathematical Intelligencer* **12**, 71-80.

Sumners, D.W., 1992, "Knot theory and DNA," in *New Scientific Applications of Geometry and Topology, Proceedings of Symposia in Applied Mathematics,* Vol. 45, D.W. Sumners (ed.), Providence, R.I.: American Mathematical Society.

Sumners, D.W., C.E. Ernst, N.R. Cozzarelli, and S.J. Spengler, 1994, "The tangle model for enzyme mechanism," (in preparation).

Wang, J.C., 1982, "DNA topoisomerases," *Scientific American* **247**, 94-109.

Wang, J.C., 1985, "DNA topoisomerases," *Annu. Rev. Biochem.* **54**, 665-697.

Wasserman, S.A., and N.R. Cozzarelli, 1985, "Determination of the stereostructure of the product of Tn3 resolvase by a general method," *Proceedings of the National Academy of Sciences USA* **82**, 1079-1083.

Wasserman, S.A., J.M. Dungan, and N.R. Cozzarelli, 1985, "Discovery of a predicted DNA knot substantiates a model for site-specific recombination," *Science* **229**, 171-174.

Wasserman, S.A., and N.R. Cozzarelli, 1986, "Biochemical topology: Applications to DNA recombination and replication," *Science* **232**, 951-960.

White, J.H., K.C. Millett, and N.R. Cozzarelli, 1987, "Description of the topological entanglement of DNA catenanes and knots by a powerful method involving strand passage and recombination," *Journal of Molecular Biology* **197**, 585-603.

Chapter 9
Folding the Sheets: Using Computational Methods to Predict the Structure of Proteins

Fred E. Cohen
University of California, San Francisco

In principle, the laws of physics completely determine how the linear sequence of amino acids in a protein will fold into a complex three-dimensional structure with useful biochemical properties. In practice, however, predicting structure from sequence remains a major unsolved problem. In this chapter the author outlines current approaches to structure prediction. The most fruitful approaches are not based on physical simulations of the folding process, but rather exploit the conservative nature of evolution. Using statistical methods, pattern matching techniques, and combinatorial problem solving, protein structure prediction is becoming steadily more tractable.

At the crossroads of physics, chemistry, biology, and computational mathematics lies the protein folding problem: How does a linear polymer of amino acids assemble into a three-dimensional object capable of executing a precise chemical function? Implicit within this question are both kinetic and thermodynamic issues: Given a particular protein sequence, what is the conformation of the folded state? What path does the unfolded chain follow to reach this folded state? This chapter outlines the history of the protein folding problem, current research efforts, the obstacles to accurate prediction of protein structure, and the areas for future inquiries.

A PRIMER ON PROTEIN STRUCTURE

Proteins are constructed by the head-to-tail joining of amino acids, chosen from a 20-letter alphabet. The 20 natural amino acids have a common backbone, but a variable side chain or R-group. The R-groups may be large or small, charged or neutral, hydrophobic or hydrophilic, and conformationally restricted or flexible (see Figure 9.1). It is the physical properties of these R-groups that determine the diverse structures into which a given amino acid chain will fold. Broadly speaking, proteins can adopt fibrous or globular shapes. Repetitive amino acid sequences adopt elongated periodic fibrous structures, with common examples including elastin (skin), collagen (cartilage), keratin (hair), and β-fibroin (silk). This chapter focuses on globular proteins.

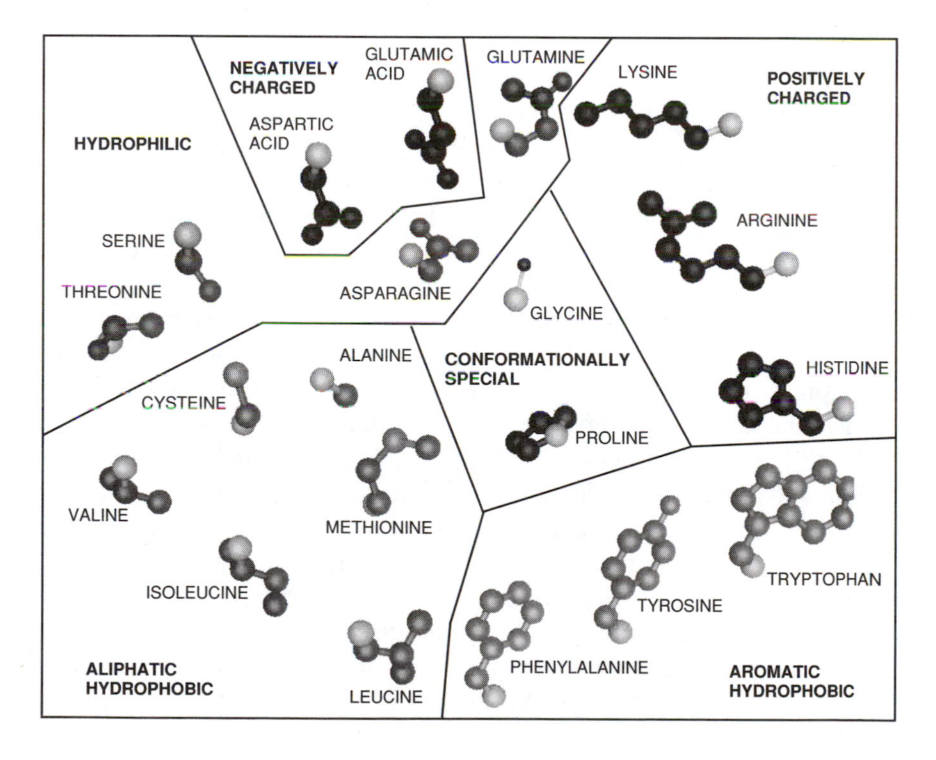

FIGURE 9.1 Twenty amino acids: R-groups are shown clustered by functional types: aliphatic hydrophobic, aromatic hydrophobic, hydrophilic, negatively charged, positively charged, and conformationally special.

The enzyme ribonuclease, which catalyzes the breakdown of ribonucleic acid (RNA), provides a useful example. The sequence contains 124 amino acids. Under appropriate conditions, the amino acid chain is covalently cross-linked in four locations through disulfide bridges between cysteines in the protein chain. (The amino acid cysteine has a reactive sulfur atom that forms such bridges, which provide the only covalent bonds joining nonneighboring amino acids in the chain.) In a classic series of experiments, Anfinsen et al. (1961) demonstrated that the amino acid sequence of ribonuclease contained enough information to code for the folded structure. Specifically, he showed that ribonuclease lost its enzymatic activity in the presence of a chemical denaturant (which disrupted the protein's structure) but spontaneously regained its activity when the denaturant was removed. Even when the disulfide pairings were scrambled after denaturation, renaturation could occur. Thus, without any outside assistance, the protein could refold. Independent of the starting conformation, the amino acid sequence contains sufficient information to direct the chain to the correct folded structure. Similar experiments have been repeated with many other proteins. This work would suggest that proteins follow an energy gradient from the denatured state to the native state. The free energy difference between these two states favors the folded state, and the height of the activation barrier along the folding pathway governs the rate of chain assembly (see Figure 9.2).

Recently, molecular biologists have discovered that some proteins can assist the folding process. These proteins, dubbed foldases, include the chaperonins (Kumamoto, 1991) that prevent proteins from assembling inside an undesirable cellular compartment, prolyl isomerases that increase the rate of the cis-trans isomerization of the amino acid proline (Fischer and Schmid, 1991), and protein disulfide isomerases (Freedman, 1989), which shuffle disulfide bridges. While it is conceivable that these foldases might take a protein to a kinetically trapped final state different from the state of lowest free energy, this seems unlikely. Instead, I imagine that these foldases simply lower the activation barrier to folding into the lowest energy state. In the absence of an appropriate foldase, the height of the activation barrier might be such that in some cases, protein folding will not occur on a biologically sensible time scale.

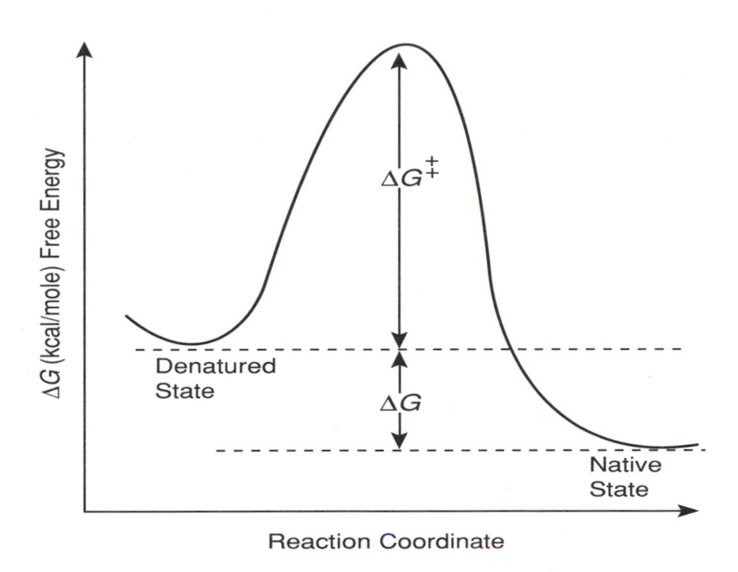

FIGURE 9.2 Thermodynamics of protein folding: the folding chain must surmount a free energy barrier ($\Delta G^{\ddagger}$) to move from the denatured to the native state. The native state is more stable than the denatured state by free energy ΔG.

One reason for the tremendous interest in the protein folding problem is that it has become simple to determine the amino acid sequence of large numbers of proteins while it remains difficult to determine the structure of even a single protein. The first protein sequences were laboriously determined by classical biochemical methods (Konigsberg and Steinman, 1977). The proteins in question were isolated, purified to homogeneity, and enzymatically digested into smaller fragments. Amino acids in each such fragment were chemically cleaved, one residue at a time, from one end and from each successive amino acid. Automated methodologies and improved chemistry accelerated this process, but protein sequencing remained a tedious task until molecular biology supplied a different approach (Maxam and Gilbert, 1980). By determining the deoxyribonucleic acid (DNA) sequence of the gene encoding the protein (using methods that were quite rapid), one could infer the amino acid sequence of the protein by simply translating the DNA codons according to the genetic code. The approach is much faster and more reliable than direct protein sequencing. With the advent of this technology has come a flood consisting of tens of thousands of protein sequences.

By comparison, the rate at which new protein structures are determined remains a trickle because the structure determination remains a formidable experimental task. X-ray crystallography was the first technique used to determine the structure of proteins (Kendrew, 1963). One must first coax a protein to crystallize with sufficient regularity to diffract X-rays. Then the crystal must be bombarded with X-rays and the X-ray diffraction pattern collected, either on film or with an electronic detector system. In principle, the X-ray diffraction pattern corresponds to a Fourier transform $\hat{D}$ of the electron density D of the crystal—with the amplitude and phase of the signal at each point corresponding to the amplitude and phase of the corresponding complex Fourier coefficient. Unfortunately, detectors can record only the amplitude, not the phase. Solving for an X-ray crystal thus involves determining the density D from $|\hat{D}|$, which can be a formidable task. In general, the problem is underdetermined. A mathematical approach is to add constraints (for example, D must be everywhere positive, since it represents a density). An experimental approach is to use additional information from the X-ray diffraction pattern obtained when the protein is crystallized in the presence of a heavy atom (for example, mercury, uranium, or platinum) or anomalous scatterers (for example, selenium) bound to the protein in a covalent or non-covalent fashion. The difference between the original and modified patterns or the patterns as a function of X-ray wavelength provides the missing phase information. Although the approach is very powerful, it requires that the protein architecture not be significantly changed by this molecular perturbation, and it is more successful when several derivatives are available for study (Blundell and Johnson, 1976). Finally, one can start with a good guess at the protein structure. The Fourier transform of this structure yields a set of intensities and phases. The hypothetical structure is rotated and translated until the intensities match the experimental data. If the correlation between the hypothetical and actual structure is strong, then the structure determination can succeed without the need for heavy atom derivatives.

More recently, nuclear magnetic resonance (NMR) spectroscopy has been used to determine protein structure (Wuthrich, 1986). Pairs of hydrogen atoms (protons) produce resonances when they lie in neighboring positions in the protein chain or when they lie very close together in space. By determining the correspondence of resonances with individual amino acids in the protein, one can determine which amino acids lie near each other. Based on these constraints, one can use the mathematical technique of distance geometry (Crippen and Havel, 1988) or restrained molecular dynamics with simulated annealing to build a partially constrained structure. (The isotopes ^{13}C and ^{15}N can also provide additional information.) Currently, this approach requires a noncrystalline but highly concentrated protein solution and works only for relatively small proteins (the resonances broaden as the molecule size increases and its tumbling time decreases).

BASIC INSIGHTS ABOUT PROTEIN STRUCTURE

If a protein sequence contains sufficient information to code for a folded structure, it should be possible to construct a potential energy function that reflects the energetics of an assembling polypeptide chain. In principle, one would "only" need to find the minimum of this potential function to know the protein's folded state. In practice, this goal has proved elusive.

Some early workers defined molecular force fields compatible with the experimentally measured conformational preferences of small molecules (Lifson and Warshel, 1969). Unfortunately, attempts to fold a denatured chain using this approach were unsuccessful (Levitt, 1976; Hagler and Honig, 1978) because multiple local minima along the potential energy surface trapped the folding chain in unproductive conformations (see Figure 9.3). Even with improved search strategies including molecular dynamics and Monte Carlo methods, it has not been possible to find the native structure from a random starting point (Howard and Kollman, 1988; Wilson and Doniach, 1989). This has been called the "multiple minima problem." It remains a critical problem for the conformational analysis of complex molecules. Despite the inability to fold proteins de novo, this approach has proved valuable for studying the behavior of proteins by studying small perturbations around the known structure.

Because direct computation is difficult, one approach would be to look for patternsand regularities in protein structures that might simplify the task of prediction. In fact, considerable insight can be gained by simply looking at experimentally determined protein structures. First of all, one

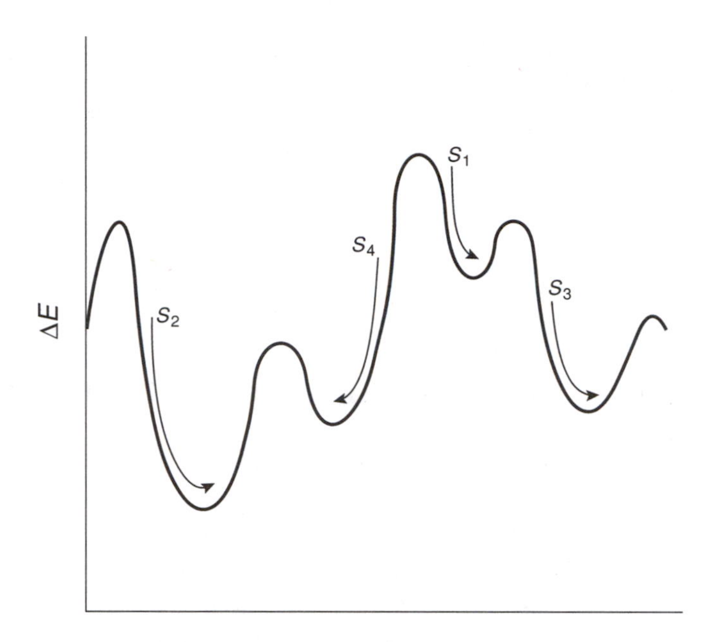

FIGURE 9.3 The multiple minima problem: a two-dimensional schematic of the energy surface of a folding protein. Different starting points lead to different metastable states. Only S_2 finds the global minimum.

observes that proteins tend to employ certain stereotypical local conformations called secondary structures. The most important are called α-helices and β-sheet structures and were suggested by Pauling (Pauling et al., 1951) based on first principles. In an α-helix, the chain follows a right-handed spiral with hydrogen bonds between the amino group (NH) of one amino acid and the carbonyl group (C=O) of an amino acid a few steps further along the chain. The result is a stable structure with a sequentially local network of hydrogen bonds (see Figure 9.4A). β-sheets offer a different solution to the hydrogen bonding problem. These sheets involve segments of the chain that are sequentially distant but conformationally similar, forming an alternating pattern of hydrogen bonds (see Figure 9.4B). The β-strands may lie parallel or antiparallel to one another. In fibrous proteins, repeated amino acid sequences yield elongated α-helices like α-keratin (or hair) and β-sheets like β-fibroin (or silk). Globular proteins must contain amino acid sequences that break α-helix and β-sheet

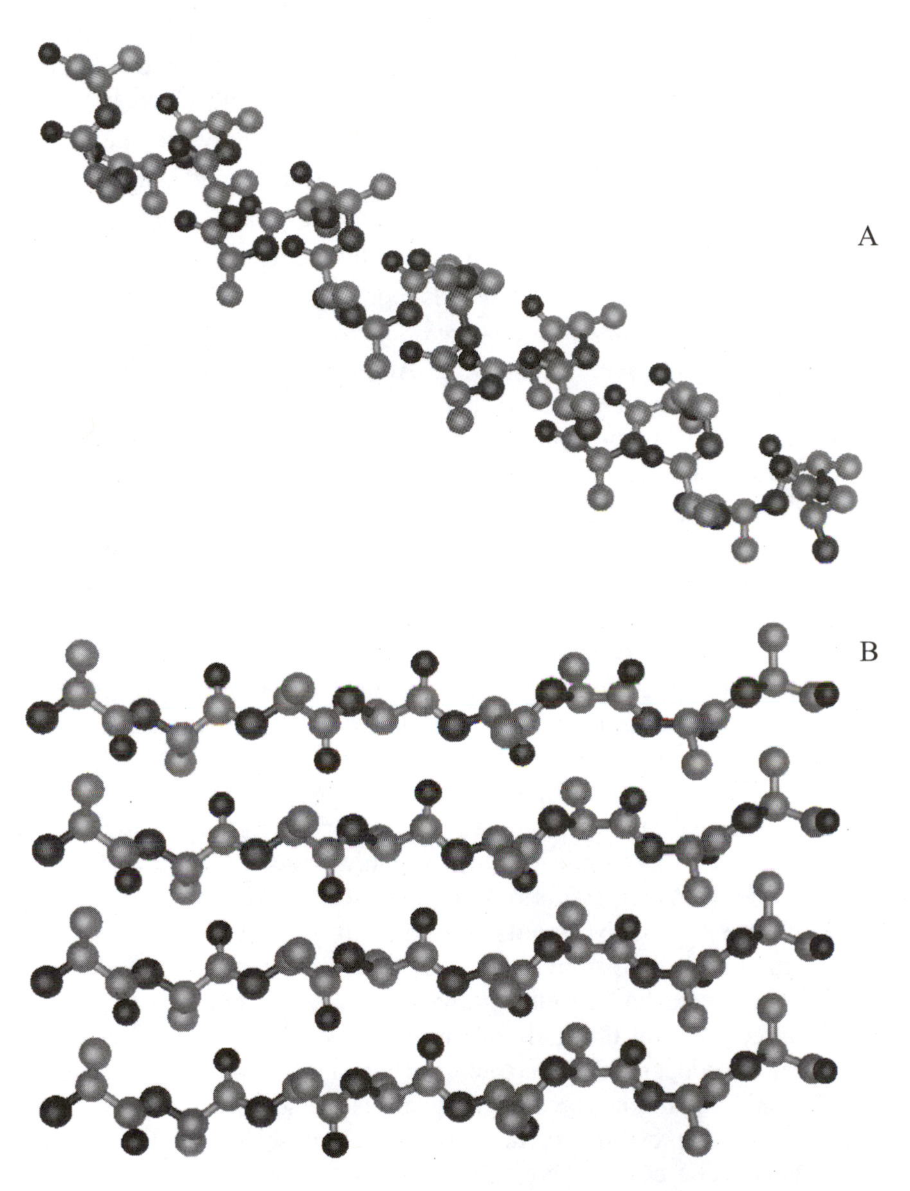

FIGURE 9.4 (A) An alpha helix. (B) A β-sheet: four parallel β-strands are shown. Hydrogen bonds exist between oxygen atoms on one strand and nitrogen atoms on the neighboring strand.

structure and cause the chain to turn back toward the center of the molecule.

Secondary structure provides a useful building block for constructing more complex protein structure (Crick, 1953; Levitt and Chothia, 1976). Proteins are usefully classified by their use of secondary structures: α/α proteins are structures dominated by α-helices (for example, myoglobin); β/β proteins are predominantly β-sheet structures (for example, plastocyanin); α/β proteins are characterized by the regular alternation of α-helices and β-strands (for example, flavodoxin); and $\alpha + \beta$ proteins are characterized by the irregular alternation of α-helices and β-strands (for example, lysozyme) (see Figure 9.5). Although the building blocks are common, the connectivity of the chain varies within these folding classes. Molecular biologists have borrowed the term "topology" (inappropriately) to describe the path that the chain takes in joining consecutive secondary structure elements. For example, many proteins contain four α-helices packed one against another to form a square four-helix bundle. With one helix taken as the reference point, the other three helices can be visited in six distinct orders. Moreover, each of these three helices can lie parallel or antiparallel to the reference helix. Thus, 48 motifs are possible. Is there any preference in the arrangements found in nature? By their general structure, α-helices have a dipole moment with partial positive charges near their N-terminus (start) and partial negative charges near their C-terminus (end). If electrostatic considerations are significant, one might expect to see antiparallel arrangements predominate (since opposite charges attract). In fact, a review of available protein structures reveals that 17 of 18 four-helix bundle structures conform to this expectation (Presnell and Cohen, 1989). Of the six possible motifs involving antiparallel arrangements, five have been observed in nature so far, and the sixth is expected to crop up as the database of protein structures grows (see Table 9.1 and Figure 9.6). An important corollary of the study of four-helix bundles is that quite distinct sequences can adopt similar structures: the code for folding is degenerate.

Further insight into protein structure is gained by considering the physicochemical properties of the different amino-acid side chains. Some side chains (those called hydrophilic) interact favorably with water, while others (called hydrophobic) do not. For globular proteins, one would expect (Kauzmann, 1959) that the hydrophilic side chains would tend to

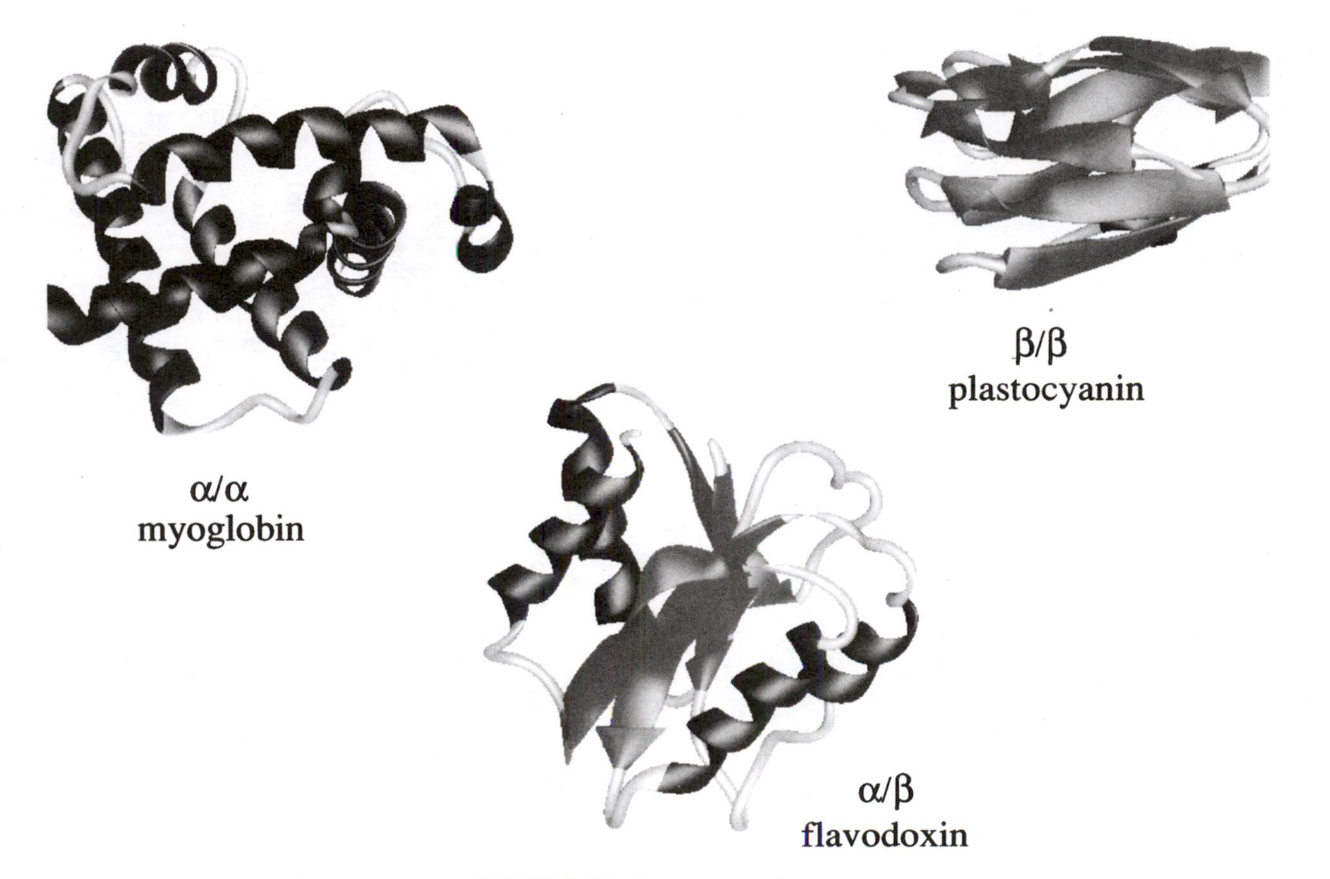

FIGURE 9.5 Tertiary structure classes.

Table 9.1 Topologies of Currently Known Four-α-Helix Bundles

Number of Overhand Connection(s)	All Antiparallel Left-handed	All Antiparallel Right-handed	Others (right-handed)
0	Complement C3a Complement C5a Cytochrome b_5 Interleukin 2 T4 lysozyme	Cytochrome *b*-562 Cytochrome *c′* Methemerythrin TMV coat protein	
1	Ferritin	Phospholipase C (b)	Cytochrome P-450_{cam}
2	Human growth hormone		

NOTE: There are no left-handed topologies for "other" four-α-helix bundles. TMV is the tobacco mosaic virus.

dominate the exterior of the protein (where it interacts with the aqueous environment) while hydrophobic side chains would occupy the molecule's interior. Richards devised a simple method for defining the "solvent-accessible" portion of a protein by rolling a sphere with a radius comparable to that of a water molecule along the molecular surface (Lee and Richards, 1971). When amino acid residues are categorized in this way, it is indeed found that hydrophobic residues tend to occur on the inside and hydrophilic residues tend to occur on the outside, although the correlation is far from perfect. Solvent-accessible surface area calculations have shed light on the importance of the "hydrophobic effect" in driving protein folding and have proved valuable in dissecting the stabilization of protein—protein and secondary structure—secondary structure inter-actions.

In summary, the analysis of protein structures has produced some unassailable conclusions: packing is an important element of protein stability; secondary structure is a common component of protein structure;

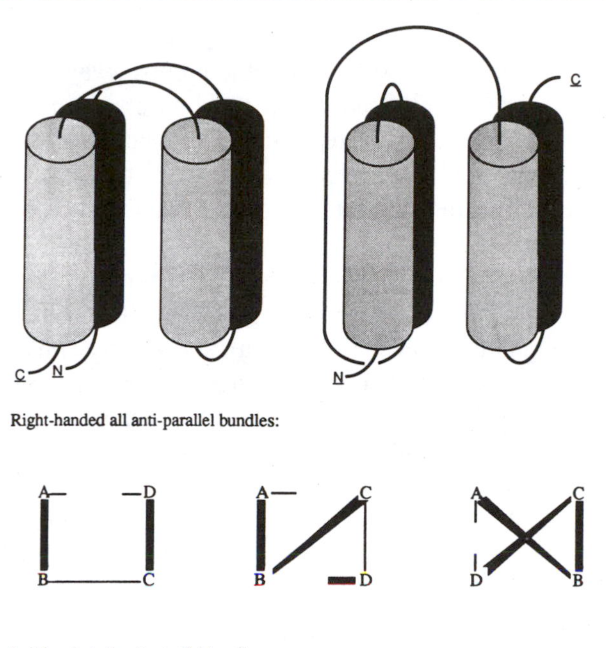

Left-handed all anti-parallel bundles:

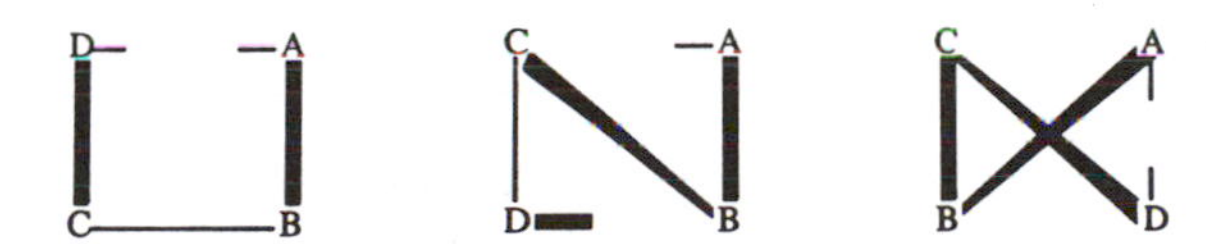

FIGURE 9.6 (Top) Two left-handed bundles (side view). Three specific attributes fully describe the topology of a four-α-helix bundle. These are the (1) polypeptide backbone connectivity between helices, (2) unit direction vectors of the individual helices, and (3) bundle handedness. In the first bundle there are no overhand connections, and in the second bundle there is one overhand connection. The handedness of a particular bundle is determined using the "right-hand rule" of physics. To determine if a helix bundle is of a particular handedness, orient the thumb of one hand parallel to the first helix or helix A where the positive unit vector stems from N-terminus to C-terminus (and helices A, B, C, and D are the first, second, third, and fourth helices on the path from the N terminus to the C terminus). Helix B should be oriented to the left if it is a left-handed bundle and to the right if it is a right-handed bundle. In the case where helix B is diagonally opposed to helix A, the handedness is based on the position of helix C relative to helices A and B. (Bottom) Schematic representation of the possible antiparallel four-α-helix bundles (top view). Bold lines represent connections in front of the page; thin lines represent connections behind the page. Left-handed and right-handed forms of four-α-helix bundles have an equal probability of occurrence. Reprinted, by permission, from Presnell and Cohen (1989). Copyright © 1989 by S.R. Presnell and F.E. Cohen.

globular proteins partition most hydrophobic groups away from the protein-solvent interface; and similar sequences yield similar structures, but quite distinct sequences can produce remarkably similar structures. Molecular biologists have suggested that nature can borrow pieces of several structures to construct new structures or simply extract a whole structure and co-opt it for a new use (Dorit et al., 1990).

This chapter focuses on two of the major computational approaches that biologists have taken to the protein folding problem: (1) Threading methods, which rely on aligning a "new" protein with sufficiently similar proteins for which structures are already known. Of course, these methods require that sufficiently similar proteins exist in the available databases. (2) Hierarchical condensation methods, which involve identification of the building blocks of protein structure and an understanding of the rules that govern the assembly of these building blocks into higher-order structure. Unlike the template methods, these methods can be applied even to proteins that do not closely resemble previously studied ones.

The rest of this chapter outlines current work, underscores the limitations of each strategy, and highlights the computational challenges we face.

THREADING METHODS

If two proteins have evolved from a common ancestor, their structures tend to remain similar (because structure underlies function) even though their sequences may diverge considerably. The converse of this statement offers a recipe for protein structure prediction: if a "new" protein sequence is recognizably similar to the sequence of a protein of known structure, then it should be possible to approximate the "new" structure by threading the sequence of the "new" protein through the "old" structure guided by the sequence alignment. The key step is to accurately align the new sequence with the known structure. While this is straightforward when the number of exact residue-for-residue matches between the two sequences exceeds 70 percent, as the degree of identity declines, this task becomes much harder (Smith and Smith, 1990). Without an accurate alignment, model-building efforts are doomed to failure.

Proteins are the translated products of information encoded in DNA stored in the chromosomes of the nucleus. As has been discussed in previous chapters, a group of three consecutive nucleotides specifies, by the genetic code, an amino acid for addition to a growing protein chain. From time to time, spontaneous mutations occur in the DNA sequence; these mutations produce changes in the amino acid sequence. While some mutations are deleterious and are eliminated by evolutionary selection, other mutations are essentially neutral in their effect and thus suffer no negative selection. As a result, the exact blueprint for a protein will drift over time as mutations accumulate. Genetic recombination can introduce additional variation to a gene. New sections of sequence can be inserted or existing sections deleted. Distinct genes or gene parts can be duplicated or concatenated. To trace the putative evolutionary relationship of two proteins, we must be able to assess the likelihood that a series of mutations, insertions, and deletions could relate the two sequences. The likelihood that any residue will be mutated into another residue could depend on the mechanisms for somatic mutation as well as on the implications of the substitution for the stability or catalytic efficiency of a protein—that is, it could depend both on mutation and selection.

We need to determine the likelihood of all possible interchanges between the 20 natural amino acids, that is, create a 20×20 transition matrix for a Markov process (although sequence evolution is not strictly Markovian). A number of approaches can be taken:

1. Focusing on mutation, we have to know the frequency of transitions ($\mathtt{A} \leftrightarrow \mathtt{G}$, $\mathtt{C} \leftrightarrow \mathtt{T}$) and transversions ($\mathtt{A} \cdot \mathtt{G} \leftrightarrow \mathtt{C} \cdot \mathtt{T}$) as well as the number of base changes required to change the triplet code for one amino acid into another. A minimum base change matrix has been constructed and used to align protein sequences. This mechanistic approach is most successful with very closely related protein sequences. Over long periods of time, however, there is ample opportunity for most mutations to occur, and so the observed spectrum of changes tends to depend more on the selective advantage or disadvantage of the amino acid substitutions than on the probabilistic nature of the mutation process itself.

2. An alternative approach to constructing the 20×20 transition matrix depends on the chemical similarity of the residues. Smith and Smith (1990) have employed this strategy to categorize and align all of the known sequences.

3. Finally, an unabashedly empirical approach can be taken. Dayhoff et al. (1972) manually aligned protein sequences from a functional family (for

example, cytochromes) and then tabulated the observed interchange frequency between all pairs of amino acids. Clearly, the success of the manual alignment and the evolutionary distance between family members have an impact on the interchange matrix. (In particular, Dayhoff's matrix corresponds to the transition matrix for the results of the Markov process over, for example, 250 million years. From this, one can infer the transition matrix for other periods.) In spite of its shortcoming, the Dayhoff matrix has proved extremely useful.

It is simplistic to expect that any one matrix can be appropriate for all positions along the chain. In fact, mutational tolerance varies widely at different positions along a protein chain (Overington et al., 1990). Regions of the protein interior and secondary structure are less tolerant of mutation. Conformational restrictions dictate the residue type in tight turns. Some cysteine side chains are covalently cross-linked to another cysteine along the chain in a disulfide bridge (and thus are under strict selective pressure to remain constant), while others exist as free sulfhydryls (and are presumably less constrained). Moreover, two cysteines in a disulfide bridge have correlated fates: if the bridge is broken by the mutation of one residue (and the protein is not sufficiently compromised that evolution selects against it), the remaining cysteine is under much less selective pressure. Proteins are highly cooperative structures, and so coupled behavior of sequentially distant residues is common. For particular proteins with well-understood structures, specific sequence "profiles" have been developed that incorporate the coupled behavior of specific amino acids along the chain. This approach performs extremely well (Bowie et al., 1991), but no general scheme exists for discerning the cooperative aspects of the sequence of a folded protein.

The existence of insertions and deletions of sequences during evolution creates a second problem in the development of an alignment metric. What is the correct penalty for creating a gap in one sequence, and how should the penalty grow as the gap becomes longer? With a sufficiently weak penalty for gaps, any two sequences can be aligned even if they are unrelated. To make matters worse, experiments have shown that the energetic penalty for inserting a sequence along the chain varies with the location of the insertion (Sondek and Shortle, 1990). Turn regions are accommodating, but the middles of secondary structure elements are more problematic. Smith has asserted that the gap penalty should be equal to

$G_1 + nG_2$, where G_1 is a penalty for creating a gap, G_2 is the penalty for extending the gap by one additional residue and n is the length of the insertion. No obvious formalism exists for determining G_1 and G_2, so they are parameters to be chosen to fit the problem at hand. Sternberg has further complicated this issue by suggesting a strategy to bias the definition of G_1 based on the position along a sequence of known structure that relies on the presence or absence of secondary structure (Barton and Sternberg, 1987). From a structural viewpoint, this is extremely sensible. From the computational point of view, the additional difficulties are considerable. More recent work by Gonnet et al. (1992) based on actual insertions and deletions suggests that gap penalties should take the form $G_1 n^{-1.7}$.

Once the mutation and gap metric have been chosen, one wishes to find the best alignment of two or more sequences. For two sequences, dynamic programming algorithms offer a rapid method for creating the optimal pairwise alignment (Smith and Waterman, 1981). For the alignment of multiple sequences, the problem becomes computationally intractable (alignment of k sequences of length N takes time $O(N^k)$), and so iterative pairwise strategies are used. The significance of a final alignment is often evaluated by comparison with a family of "random" sequences. Unfortunately, protein sequences are far from random, and so this approach usually overestimates the significance of an alignment. New methods are being developed to improve the scoring metrics and to produce "random" sequences with the Markov dependences commonly observed in proteins (Karlin and Altschul, 1990).

When an accurate alignment is available for a new protein sequence and a protein of known structure, it is possible to construct a useful model of the "new" three-dimensional structure based on the "old" structure. For closely related structures, the residue-by-residue positional error is small. However, the error grows as the divergence between the sequences of the model and known structures increases (Chothia and Lesk, 1986). Because "reasoning by homology" plays such an important role in the prediction of protein structure, various recipes have been developed for model building (Greer, 1990; Chothia et al., 1989; Jones and Thirup, 1986; Bruccoleri and Karplus, 1987; Ponder and Richards, 1987; Wilson et al., 1993). One method is based on using secondary structure segments to create a molecular scaffold, thereby decomposing the problem. Onto this scaffold, particular loops can be added, and the side chain conformation of the individual residue can be specified. The loops can be selected from a "loop thesaurus" (although this suffers from the drawback that it cannot predict novel loops) or by computational search (although this becomes prohibitive for loops of more than 10 residues). The conformation of the side chains cannot be determined by direct conformational search because too many local minima exist, but the problem can be made more tractable by focusing the computation on "rotamer libraries" containing the most commonly found conformations and by performing comparisons with the conformations found in collections of related sequences (Wendoloski and Salemme, 1992).

While model construction is straightforward, model validation is much more difficult. Is a homology-built model correct? If not, where is it wrong, and how can it be improved? The problem turns out to be surprisingly hard. Indeed, Novotny et al. (1988) showed that it is possible to construct "incorrect" models that nonetheless satisfy a variety of energetic constraints. They took hemerythrin and an immunoglobulin domain, two proteins with entirely different secondary and tertiary structure. The hemerythrin sequence was forced into the immunoglobulin structure and vice versa. The difficulty in distinguishing the correct and misfolded structures based on energy calculations was both remarkable and profound. This suggests that the errors inherent in existing potential functions may dominate the energy differences between correctly and incorrectly folded proteins. In practice, homology modeling has had varied success (Pearl and Taylor, 1987; Wierenga and Hol, 1993), ranging from important triumphs to ambiguous or erroneous results.

Clearly, better methods are needed to evaluate the quality of model-built structures. One such test stems from the observation that a hallmark of folded proteins is the close packing of atoms (Richards, 1977). Using various computational methods, various workers (Richards, 1977; Frauenfelder et al., 1987) have shown that the packing density within a protein turns out to be comparable to that seen in the crystalline arrangement of small organic molecules, with cavities being rare (and those that are observed can be functionally relevant). Gregoret and Cohen (1990) have exploited tests for packing defects as a strategy for sorting correct structures from their incorrectly modeled counterparts. While this approach provides a useful filter for rejecting some structures, well-packed but incorrect structures are found.

A second approach to distinguishing correct from incorrect structures is based on the concept of solvent-accessible surface area, a notion that has found broad utility in the analysis of the energetics of protein folding and molecular recognition (Lee and Richards, 1971). The solvent-accessible surface area is defined as the area of the locus traced by the center of a sphere of radius 1.4 angstroms (Å) (the radius of a water molecule) when it is rolled along the surface of the protein. The solvent-accessible contact area is defined as the region of the molecule that contacts the probe sphere. These two measures tend to be roughly proportional and can be calculated using numerical integration (Lee and Richards, 1971) or analytical methods from differential geometry (Richmond, 1984). Solvent accessibility calculations turn out to be quite relevant to the energetics of protein folding (Chothia, 1974). While side chains on the exterior of a protein are in contact with water, the interior side chain environment resembles the organic solvent octanol (an eight-carbon alcohol). It is this difference that makes it favorable for hydrophilic side chains to lie on the surface and hydrophobic residues to cluster in the interior. The free energy required to transfer particular amino acid side chains from octanol to water has been measured (Lesser and Rose, 1990). The experimental values correlate closely with the accessible surface area of the side chain (1-$Å^2$ surface area = 47 calories per mole), and the data provide a basis for distinguishing correct from incorrect protein structures. Indeed, Novotny et al. (1988) found that it was possible to sort the correct hemerythrin and immunoglobulin structures from their incorrect models by taking into account the accessible surface area. For this and other reasons, fourth-generation potential functions are being developed that include a solvent accessibility term as a method for implicitly incorporating protein-solvent interactions.

Another issue in evaluating proposed protein structures is the treatment of electrostatic interactions (Gilson and Honig, 1986). Proteins contain many charged side chains, and the peptide backbone forms a dipole. The dielectric behavior of the protein interior is significantly different from the behavior at or near the protein-solvent interface. This creates difficulties in computational efforts to evaluate properties of proteins. For example, a simple coulombic formulation of electrostatic interactions does a poor job of replicating experimental data on the intrinsic affinity of an ionizable group for a proton (pKa). Since the groups in the active site of a protein often have unusual pKa's, advances are necessary to improve our understanding of catalysis. Honig and co-workers have used the Poisson-Boltzmann equation to obtain a better understanding of protein electrostatics (Gilson and Honig, 1986). Distinct dielectric environments can be accommodated, and while closed-form solutions are not possible for any complex systems, finite difference methods can be used to calculate the field strength. These studies have proved useful in replicating pKa data, and efforts are under way to incorporate the Poisson-Boltzmann formalism into existing potential functions. Unfortunately, these calculations are computationally intensive. Little has been tried to exploit electrostatic interactions as a guide to the merits of a protein structural model.

Finally, experimental information can be used to sort correct from incorrect structures. The relative proportion of α-helix and β-sheet in a protein can be measured by circular dichroism spectroscopy. Discrepancies between the observed and the predicted data can argue against a model structure. The precise cross-linking of the polypeptide chain through disulfide bridges or other chemical reagents provides distance constraints that connect sequentially distant regions of the chain. To a lesser extent, site-directed mutagenesis, which detects the impact of changes to the amino acid sequence on protein stability and function, and limited proteolysis experiments, which detect the relative accessibility of regions of the chain, can provide useful constraints to sort between alternative model structures. The combination of theoretical methods and low-resolution experimental tools holds great promise for directing the construction of useful protein structure models.

PREDICTING HIV PROTEASE STRUCTURE: AN EXCURSION

Several types of proteases, enzymes that catalyze the breakdown of proteins, have been characterized by biochemists. Frequently, they are named after a key component of their catalytic machinery: serine, cysteine, aspartyl, or metallo proteases. With the discovery of the AIDS virus (HIV) and the determination of its genomic sequence, it appeared that the virus produced a great deal of its enzymatic and regulatory machinery as one long incapacitated polyprotein that required a specific protease to split it into active components. Sequence analysis suggested a region of the genome that could code for a protein with some resemblance to the aspartyl proteases known from prokaryotic and eukaryotic sources. The problems were that the sequence was too short (~ 100 residues) when compared to the known aspartyl proteases (~ 240 amino acids) and that it contained only one of the two key aspartyl groups that form the active site. Moreover, the degree of sequence similarity between the viral sequence and the known aspartyl proteases was sufficiently low that researchers were unsure of exactly what went where. This was of more than passing interest as genetic studies of HIV demonstrated that mutation to the putative aspartyl protease blocked viral replication, and hence this was a promising target for pharmaceutical intervention.

Pearl and Taylor (1987) studied the structures of several aspartyl proteases and constructed a template that encoded the essential features of this class of enzymes. A sequence template was found that could be used to scan the database of known sequences and efficiently sort aspartyl proteases from all other proteins. The HIV sequence fit half of the template and exhibited very economical tendencies with regard to the loop regions that joined the β-strands in the molecular framework. The prokaryotic and eukaryotic aspartyl proteases contain two domains that are structurally similar and can be related by a dyad axis. Moreover, one of the two aspartate residues in the active site is contributed by each domain. Pearl and Taylor reasoned that HIV, in an attempt to achieve additional genomic economy, elaborated a protease that had to pair or dimerize to form the active enzyme. A three-dimensional model of the HIV protease was constructed by following the template-directed homology to the aspartyl proteases of known structure that facilitated subsequent chemical and biochemical efforts. Their structural model proved to be extremely insightful when compared with the structural data provided by X-ray crystallographers several years later.

HIERARCHICAL APPROACHES

If protein folding is so hard, how do proteins manage to get it right? Does the folding process obey simple, logical rules that could guide computational efforts to reproduce it? Biochemists tend to describe protein structure in a hierarchical fashion, and many believe that the folding process tends to proceed up the hierarchy (Linderstrom-Lang and Schellman, 1959). According to this view, the amino acid sequence or primary structure would first collapse from a disordered chain to form ordered elements of secondary structure, α-helices and β-strands. These secondary structure elements would coalesce to form a stable tertiary structure, consisting of the packing of the secondary structure elements against one another in the complete protein molecule. Finally, individual protein monomers with stable tertiary structures can sometimes aggregate to form multimers with the interaction defining the quaternary structure, often having complex functional and regulatory roles. This suggests a computational strategy to relate protein sequence and structure. First, predict the location of α-helices and β-strands, and then pack secondary structure units together to form an approximate tertiary structure that can be refined to the folded protein structure (see Figure 9.7).

Is secondary structure prediction possible? Stated more precisely, is secondary structure determined predominantly by local interactions along the chain, or is it dependent on numerous nonlocal (tertiary) interactions? Various small peptides have been synthesized and spectroscopically characterized in an effort to understand the origins of the stability of an α-helix (Padmanabhan et al., 1990). Experiments show that several short sequences (< 20 residues in length) can form stable α-helices and the individual amino acid conformational preferences correlate with those observed in globular proteins. Studying β-structure in this fashion, however, has proved more difficult.

Studies of whole proteins shed further light on the degree to which secondary structure arises from local interactions. Experiments have suggested that one can identify a "molten globule" state that can be stabilized under acidic conditions (Kuwajima, 1989). This intermediate state appears to contain native-like secondary structure as inferred from circular dichroism studies, but the tertiary structure appears not to be present. In short, secondary structure appears to be stable in the absence of tertiary or nonlocal interactions. On the other hand, the local information producing secondary structure cannot be *too* local. For it is known that identical pentapeptide sequences chosen from distinct proteins can adopt entirely different structures (Kabsch and Sander, 1984). For example, the same pentapeptide may form part of an α-helix in one protein and a β-strand in another. Thus, the necessary information must extend beyond

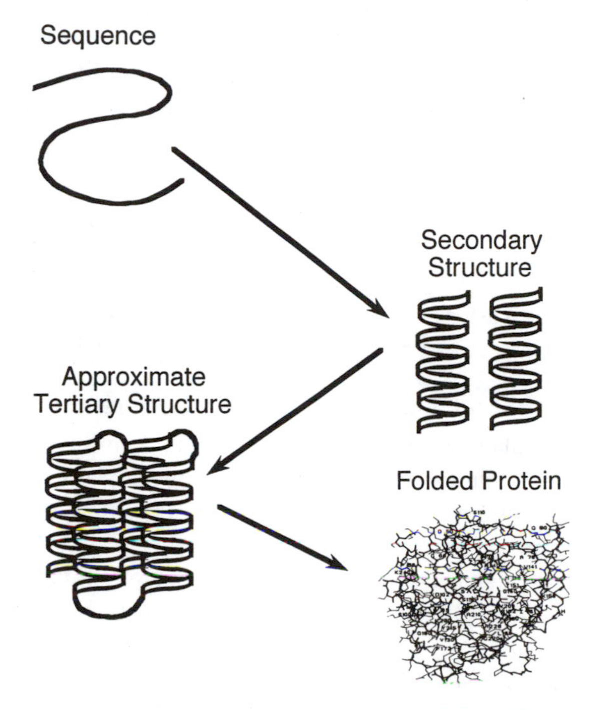

FIGURE 9.7 A hierarchical condensation model for protein folding. Sequence determines secondary structure, and secondary structure elements assemble to form an approximate tertiary structure. Energy refinement yields a detailed three-dimensional structure.

five residues. Presumably, the solution is that the conformation of some sequences is specified in large part by local interactions, while others are stable only in the context of the neighboring sequences. The most difficult challenge for secondary structure prediction methods is to determine the structure of these context-dependent regions.

Two general strategies have been applied to the secondary structure prediction problem: statistical approaches and rule-based approaches.

The statistical approaches assert that proteins of known tertiary structure provide a useful data set describing secondary structure preferences of individual amino acids. Two presumptions are made: tertiary structure will exert no consistent effects on secondary structure, and the existing database is of sufficient size to provide important information. The first assertion recalls our discussion of the local versus global determinants of protein organization. I believe that local nteractions play a significant role in protein folding, but, in the literature, this remains an open question. The adequacy of our current database can be approached in a more straightforward way. The conformations of over 40,000 amino acids in approximately 200 distinct protein structures are known. These distribute between α-structure (~ 30 percent), β-structure (~ 30 percent), and turns or coils (~ 40 percent). The likelihood that alanine will appear in an α-helix ($L_\alpha(\text{Ala})$) can be calculated easily from this data set.

$$L_\alpha(\text{Ala}) = \frac{\text{number of alanines in } \alpha\text{-helices}}{\text{number of alanines in data set}}.$$

Even the 400 amino acid doublet propensities, which reflect the conditional probability that an alanine will occur in an α-helix contingent on the amino acid type of the neighboring residue, can be usefully approximated. However, the current data set is not adequate to provide information about the 8,000 triplet amino acid preferences. Moreover, it is not clear that the triplet interaction preferences will be the sum of three doublet interactions or that complete triplet preferences adequately define the conformational preferences of amino acids. Additional protein crystal structures and studies on model peptide systems will help in overcoming these limitations.

In 1974, a landmark paper on protein secondary structure prediction was published by Chou and Fasman (1974). Working with a much smaller protein database, the authors calculated the secondary structure propensities of each amino acid, for example,

$$P_\alpha(\text{Ala}) = \frac{\text{number of alanines in } \alpha\text{-helices}}{\text{number of alanines not in } \alpha\text{-helices}}.$$

From this information, residues were classified as helix formers ($P_\alpha \geq 1.05$), intermediate ($0.70 \leq P_\alpha < 1.05$), and helix breakers ($P_\alpha < 0.70$). Local clusters of helix formers defined helical nucleation sites. These nuclei were extended toward the N- and C-termini following rules based on the aggregate P_α's. Although no computer algorithm accompanied the initial work, the method was sufficiently simple that it could be applied by hand. The accuracy of this algorithm (that is, the percentage of correct predictions on a residue-by-residue basis) approached 60 percent.

Subsequent work has developed more sophisticated variations on the theme. In 1979, Robson and co-workers introduced an information-theoretic formalism to supplement conformational preferences of the isolated residues with preferences based on pairwise interactions (Garnier et al., 1978). The method was easy to implement in a computer algorithm and achieved ~ 64 percent accuracy. More recently, various authors (Qian and Sejnowski, 1988; Holley and Karplus, 1989) have employed neural networks, which belong to a general class of machine learning algorithms that can efficiently "learn" an optimal translation of one data string (for example, a protein sequence) into another (for example, the sequential secondary structure assignments). The network is a group of input nodes connected to a group of output nodes with an optional hidden layer or layers of nodes (see Figure 9.8). A matrix of weights is developed to map the input information into the nodes on a path to the output layer. Like neurons in the nervous system, a cooperative nonlinear "firing" potential is used to decide if adequate information has accumulated to switch on an output node (see Figure 9.9). For secondary structure prediction, this "all or none" output node predicts an α-helix when the accumulated helical propensity of the residue of interest and its neighbors crosses the threshold. The weights for the connections that relate input nodes to output nodes are learned by example. A window specifies the number of neighboring residues that can contribute to the conformational state of the residue of interest. Case after case of input amino acid sequence and output secondary structure is presented to the network. A least squares algorithm defines an optimal set of weights for the encoding of the data set using a back propagation strategy. A more complete description of neural networks can be found in a chapter of the book *Parallel Distributed Processing: Explorations in the Microstructure of Cognition,* Vol. 1 (Rumelhart et al., 1986).

Neural networks easily achieve an accuracy of 64 percent, a figure comparable to that for other methods. It is useful to explore the connection weights derived by the network that relate amino acids to their secondary structure preferences. Figure 9.10 is a Hinton diagram of these weights (the magnitude of the weight is proportional to the area of the square; positive weights are in white, and negative weights are in black). Alanine

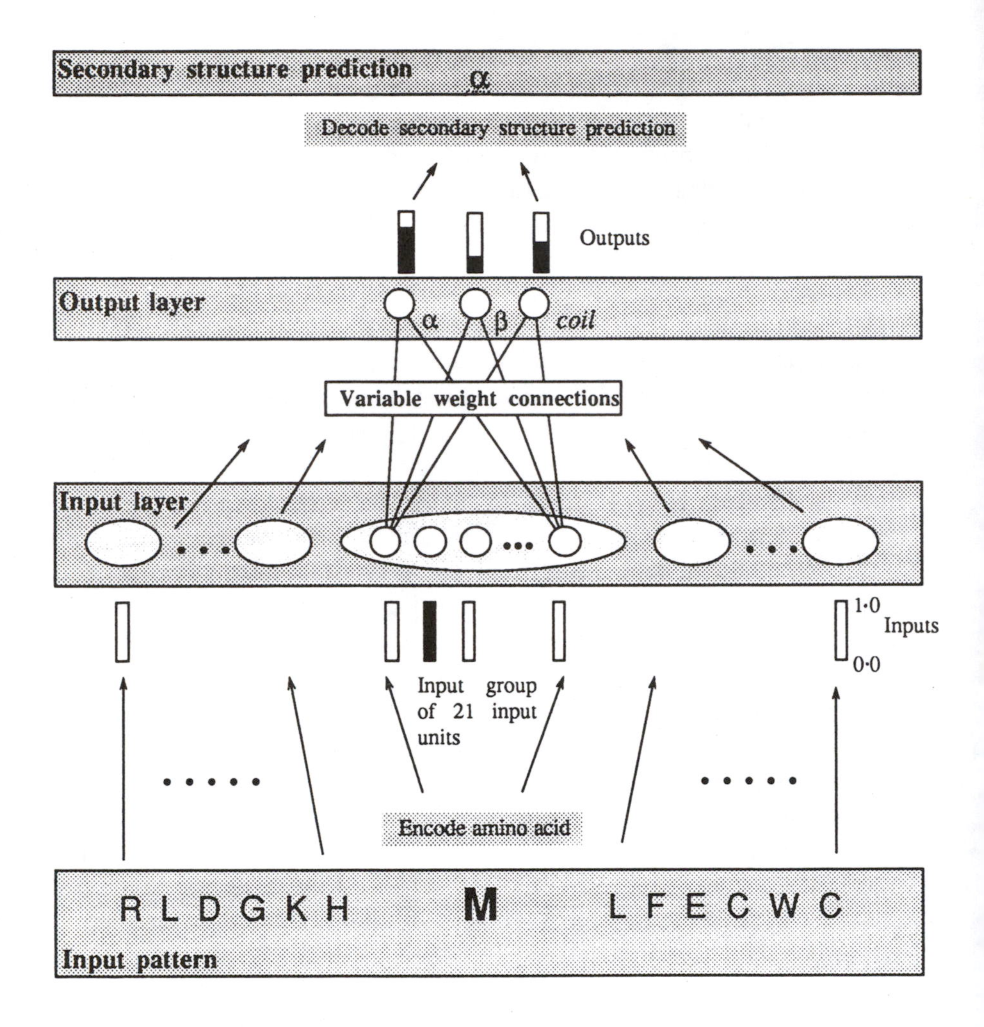

FIGURE 9.8 The secondary structure neural network. The input pattern is a sequence of amino acids centered around a central amino acid. Each amino acid is mapped to one input group, which is a collection of 21 units. Each amino acid causes an input of 1.0 to one of the units of its group and an input of 0.0 to the other units. We typically use the six amino acid residues on each side of the central amino acid, for a total of 13 × 21 = 273 input units. There are three output units: helix, strand, and coil. Each input unit is connected to each output unit, and the output unit with the greatest output is taken to be the secondary structure prediction for the central amino acid. Additional input units can be accommodated. Reprinted, by permission, from Kneller et al. (1990). Copyright © 1990 by Academic Press Limited.

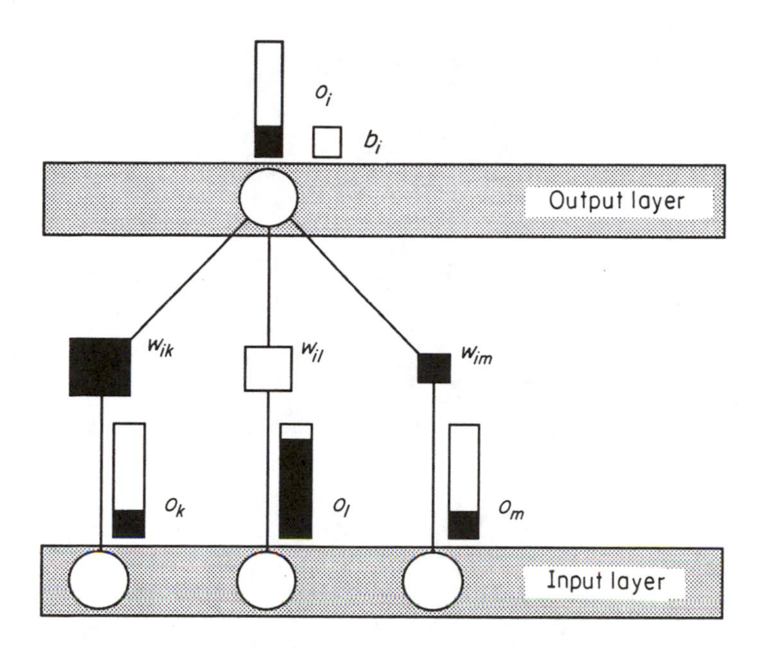

FIGURE 9.9 The basic neural network. Circles represent the units, and squares the weights between the units. The larger the square, the greater its absolute value. Solid squares represent negative values, and open squares represent positive values. The bars represent the outputs of the units, with values ranging from 0.0 to 1.0. The symbols are defined as follows: o_k is the output of unit k; w_{ik} is the weight to unit i from unit k; and b_i is the bias of unit i. The activation of unit i is $a_i = \Sigma w_{ik} o_k + b_i$, and its output is $o_i = (1 + e^{-a_i})^{-1}$. Reprinted, by permission, from Kneller et al. (1990). Copyright © 1990 by Academic Press Limited.

and leucine are seen to strongly favor α-helices, while proline and glycine disrupt the α-helix structure and prefer turn conformations. This is consistent with the structural information derived from previous studies and reinforces the sensibility of this approach.

Why do neural networks not perform more accurately? We have begun to address this question. If a network is trained on proteins restricted to one structural class, especially all helical proteins, the accuracy improves significantly (Kneller et al., 1990). For example, nodes can be added that capture the alternating distributions of hydrophobic and

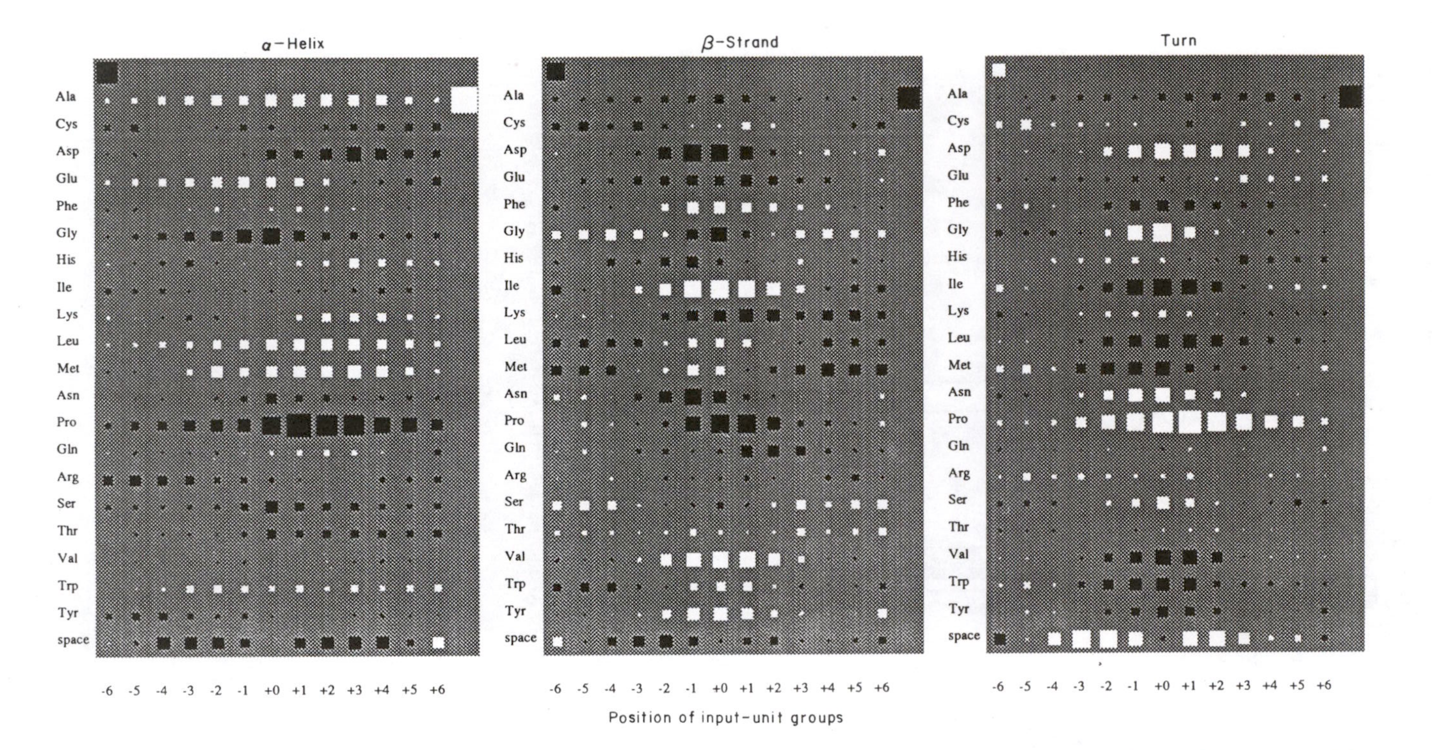

FIGURE 9.10 Hinton diagram of the weight matrix. The weights connecting each input unit to each output unit are shown. The size of each box correlates with the magnitude of the weight. Solid boxes have negative values, and open boxes have positive values. The three output units are shown as separate blocks. The 273 input units connected to each output unit are divided into 13 groups of 21 units each. Along the abscissa, the groups are located at positions in the range $-6, -5, \ldots, 0, \ldots, 5, 6$ relative to the central residue. Along the ordinate, the 21 units are labeled by residue type. Reprinted, by permission, from Kneller et al. (1990). Copyright © 1990 by Academic Press Limited.

hydrophilic residues in a phase-insensitive way. Together, these methods improve secondary structure prediction accuracy for all helical proteins to 79 percent and for all β proteins to at least 70 percent. α/β proteins remain problematic. Presumably, this relates to the fact that the fundamental structural unit in α/β proteins involves an α-helix and the preceding and/or following β-strands. This super-secondary structure involves at least 30 residues, far more than are included in the windows currently used. By exploiting a family of aligned protein sequences, improvements in secondary structure prediction are anticipated. Other aspects of this problem continue to make this a fertile area for study.

The second general approach to secondary structure prediction is rule-based methods, which try to capture biochemical regularities in protein structure. In an important early paper, Lim (1974) described "rules" that specify residue combinations along the chain that stabilize or destabilize α-helices and β-sheets. The rules attempted to capture the notion that secondary structure elements need to be compatible with the overall tertiary structure consisting of a hydrophobic core and hydrophilic exterior. Among other constraints, isolated β-strands are stable only in the context of larger β-sheets, and the edge strands in these β-sheets have significantly different properties than the interior strands. Technical difficulties in the formulation of the rules hampered efforts to implement these ideas in a computer program, but this does not detract from the insightfulness of the approach.

Our group has followed up on the rule-based approach pioneered by Lim. We have constructed PLANS, a Pattern Language for Amino and Nucleic Acids Sequences, and implemented this language in LISP and C (Cohen et al., 1983, 1986). Accurate patterns can be written to identify various structural features. For example, rule-based patterns can be used to identify "turns" in protein chains. Turns contain hydrophilic stretches without periodic structure; they tend to be evenly distributed throughout the protein chain. Extremely hydrophilic clusters of amino acids are nearly always turns. Weaker turns can be identified as relatively hydrophilic clusters of residues appropriately separated from the more obvious turns. The spacing between turns depends on the secondary structure content of the protein. For example, an α-helical segment bounded by turns contains approximately twice as many residues as a similar β-strand segment. The class of a protein (α/α, α/β, β/β) offers a simple way of specifying the expected link length between turns. Using a hierarchical set of turn patterns embodying these ideas, one can identify turns with ~ 90 percent accuracy.

Other work on rule-based methods has focused on finding the exact location of α-helices on α/α proteins (Presnell et al., 1992). Even though helices are heterogeneous objects, patterns have been developed to recognize their beginnings (N-caps or N-termini), cores, and ends (C-caps or C-termini). While the core patterns are very accurate (> 90 percent of helices can be located), the termini, especially the C-termini, remain poorly defined. Because of these deficiencies, amalgamation of the three groups of patterns produces a secondary structure prediction that is only 71 to 78 percent correct. Developing sequence patterns to represent protein substructures is labor intensive. Recently, there have been attempts to automate this process by means of heuristic, iterative algorithms for pattern construction (King and Sternberg, 1990).

The hierarchical approach to protein structure prediction is premised on the notion that secondary structure will be a useful computational intermediate for the prediction of overall tertiary structure. How exactly can one use secondary structure information to bootstrap the process? Conceptually, the most straightforward approach to this problem would be to construct all possible three-dimensional arrangements of the secondary structure segments and then eliminate those with structural flaws (high potential energy).

Combinatorial approaches can be used to search over the many possible arrangements and evaluate their plausibility (Cohen et al., 1979; Ptitsyn and Rashin, 1975). The approach is particularly well developed for the case of α-helices, owing to the fact that the periodicity of α-helices tends to favor distinct packing geometries for pairs of α-helices (Crick, 1953; Chothia et al., 1977; Richmond and Richards, 1978; Murzin and Finkelstein, 1988). Hydrophobic residues tend to dominate the interfacial region between paired α-helices. Moreover, there is a correlation between the extent of the hydrophobic interface on the α-helices and the preferred packing geometry. In the next section, we describe an application of this approach to the oxygen-bearing protein myoglobin. Similar work has been done on β/β (Cohen et al., 1980, 1982; Finkelstein and Reva, 1991). For α/β proteins, the combinatorial complexity of these proteins is much greater than for β/β and α/α proteins, but it is still possible to use secondary structure as a guide to approximate tertiary structure. A complete description of these combinatorial algorithms can be found in a recent review (Cohen and Kuntz, 1989).

If the hierarchical approach to protein structure prediction is to succeed, secondary structure prediction must improve (to at least the 80 percent accuracy level), the combinatorial methods must be further refined, and the radius of convergence of existing potential functions must be extended to allow optimization of the final structure (Wilson and Doniach, 1989; Sippl, 1990; Troyer and Cohen, 1991).

PREDICTING MYOGLOBIN STRUCTURE: AN EXCURSION

Myoglobin, a 153-residue oxygen-carrying protein, was the first protein structure to be determined by X-ray crystallography. It is composed of six long α-helices and two other smaller helices that do not contribute to the protein's hydrophobic core. In the 1970s we showed that it is possible to construct three-dimensional models of myoglobin by the successive addition of helices to an initial helix using the putative hydrophobic interfaces, while respecting the geometric preferences of helix-helix interactions. From the work of Pauling (1967), we know the conformation and hence the atomic coordinates of the backbone of an α-helix. To begin, we are free to place helix A (residues 3 through 18) so that its axis is coincident with the x-axis with its centroid at the origin. Residue 10 is the center of a potential helix-helix interaction site and creates a sticky patch on the surface of helix A. One possible pairing of helices would join A and B (20 through 35) through sticky patches centered at residues 10 and 28. A line segment perpendicular to the axis of helix A that passes through the C_α of residue 10 with a length of 8.5 Å can be used to place helix B such that the segment passes through the C_α of residue 28, is normal to the axis of helix B, and terminates at this axis. Helix E could be placed via its interaction with helix B, and so on. While the packing of helices B and E will be sensible, nothing in this procedure prevents helices A and E from colliding. For the six helices of myoglobin, there are 14 likely helix-helix interaction sites and 3.4×10^8 possible structures.

To complete this calculation, a PDP 11/70 filling an entire machine room used to work for 48 hours. Today, a laptop computer can complete this same calculation in much less than one hour. An algorithm with a tree architecture can be used to generate these structures. Fortunately, many of the possible structures violate steric constraints (that is, parts of the molecule collide) or disrupt the connectivity of the chain (that is, the interhelix portion of the protein chain cannot stretch from the end of one helix to the start of the next helix), and so large branches of the "tree" can be removed from further consideration. Remarkably, only 20 plausible structures are obtained. Using the additional information that myoglobin contains an iron-bearing heme group, the list can be winnowed: only 2 of the 20 structures could use two histidines to chelate an iron atom surrounded by a heme ring in a sterically reasonable way. As it happens, these 2 structures are extremely similar (root-mean-square displacement (rmsd) between C_α atoms = 0.7 Å) and resemble the crystal structure of myoglobin (rmsd = 4.4 Å). Presumably, detailed energy calculations could be used to refine these structures. To date, the radius of convergence of existing molecular dynamics algorithms is too small to close the 4-Å gap between these approximate structures and the X-ray structure.

CONCLUSION

The protein folding problem is enormously important to biologists. Sequences for exciting new proteins are relatively easy to determine. Structural data for these molecules are much more difficult to obtain. Yet proteins contain a structural blueprint within their sequence. The computational challenge to unravel this blueprint is great. This chapter has highlighted the important problems in this field and identified fertile territory for new investigations.

ACKNOWLEDGMENTS

This work was supported by grants from the National Institutes of Health, the Searle Scholars Program, and the Advanced Research Projects Agency.

REFERENCES

Anfinsen, C.B., E. Haber, M. Sela, and F.H. White, 1961, "The kinetics of the formation of native ribonuclease during oxidation of the reduced polypeptide domain," *Proceedings of the National Academy of Sciences USA* **47**, 1309-1314.

Barton, G.J., and M.J.E. Sternberg, 1987, "Evaluation and improvements in the automatic alignment of protein sequences," *Protein Engineering* **1**, 89-94.

Blundell, T.L., and L.N. Johnson, 1976, *Protein Crystallography,* New York: Academic Press.

Bowie, J.U., R. Luthy, and D. Eisenberg, 1991, "A method to identify protein sequences that fold into a known 3-dimensional structure," *Science* **253**, 164-170.

Bruccoleri, R.E., and M. Karplus, 1987, "Prediction of the folding of short polypeptide segments by uniform conformational sampling," *Biopolymers* **26**, 137-168.

Chothia, C., 1974, "Hydrophobic bonding and accessible surface area in proteins," *Nature* **248**, 338-339.

Chothia, C., and A.M. Lesk, 1986, "The relation between divergence of sequence and structure in proteins," *EMBO J.* **5**, 823-826.

Chothia, C., A.M. Lesk, A. Tramontano, M. Levitt, S.J. Smith-Gill, G. Air, S. Sheriff, E.A. Padlan, D. Davies, and W.R. Tulip, 1989, "Conformations of immunoglobulin hypervariable regions," *Nature* **343**, 877-883.

Chothia, C., M. Levitt, and D. Richardson, 1977, "Structure of proteins: Packing of α-helices and pleated sheets," *Proceedings of the National Academy of Sciences USA* **74**, 4130-4134.

Chou, P.Y., and G.D. Fasman, 1974, "Conformational parameters for amino acids on helical, β-sheet, and random coil regions calculated from proteins," *Biochemistry* **13**, 211-245.

Cohen, F.E., M.J.E. Sternberg, and W.R. Taylor, 1980, "Analysis and prediction of protein β-sheet structures by a combinatorial approach," *Nature* **285,** 378-382.

Cohen, F.E., M.J.E. Sternmerb, and W.R. Taylor, 1982, "The analysis and prediction of tertiary structure of globular proteins involving the packing of α-helices against a β-sheet: A combinatorial approach," *Journal of Molecular Biology* **156**, 821-862.

Cohen, F.E., R.A. Abarbanel, I.D. Kuntz, and R.J. Fletterick, 1983, "Secondary structure assignment for α/β proteins by a combinatorial approach," *Biochemistry* **22**, 4894-4904.

Cohen, F.E., R.A. Abarbanel, I.D. Kuntz, and R.J. Fletterick, 1986, "Turn prediction in proteins using a pattern-matching approach," *Biochemistry* **25**, 266-275.

Cohen, F.E., and I.D. Kuntz, 1989, "Tertiary structure predictions," pp. 647-706 in *Prediction of Protein Structure and the Principles of Protein Conformation,* G.D. Fasman (ed.), New York: Plenum.

Cohen, F.E., T.J. Richmond, and F.M. Richards, 1979, "Protein folding: Evaluation of some simple rules for the assembly of helices into tertiary structures with myoglobin as an example," *Journal of Molecular Biology* **132**, 275-288.

Crick, F.H.C., 1953, "The packing of α-helices: Simple coiled coils," *Acta Crystallogr.* **6,** 689-697.

Crippen, G.M., and T.F. Havel, 1988, *Distance Geometry and Molecular Conformation,* New York: John Wiley & Sons.

Dayhoff, M.O., L.T. Hunt, P.J. McLaughlin, and D.D. Jones, 1972, "Gene duplications in evolution: The globins," pp. 17-30 in *Atlas of Protein Sequence and Structure,* Vol. 5, M.O. Dayhoff (ed.), Silver Spring, Md.: National Biomedical Research Foundation.

Dorit, R.L., L. Schoenbach, and W. Gilbert, 1990, "How big is the universe of exons?" *Science* **250**, 1377-1382.

Finkelstein, A.V., and B.A. Reva, 1991, "A search for the most stable folds of protein chains," *Nature* **351**, 497-499.

Fischer, G., and F.X. Schmid, 1991, "The mechanism of protein folding: Implications of in vitro refolding mode for de novo protein folding and translocation in the cell," *Biochemistry* **29**, 2205-2212.

Frauenfelder, H., H. Hartmann, M. Karplus, I.D. Kuntz, Jr., J. Kuriyan, F. Darak, G.A. Petsko, D. Ringe, R.F. Tilton, Jr., and M.L. Connolly, 1987, "Thermal expansion of a protein," *Biochemistry* **26**, 254-261.

Freedman, R.B., 1989, "Protein disulfide isomerase: Multiple roles in the modification of nascent secretory proteins," *Cell* **57**, 1069-1072.

Garnier, J., D.J. Osguthorpe, and B. Robson, 1978, "Analysis of the accuracy and implications of simple methods for predicting the secondary structure of globular proteins," *Journal of Molecular Biology* **120**, 97-120.

Gilson, M.K., and B.H. Honig, 1986, "The dielectric constant of a folded protein," *Biopolymers* **25**, 2097-2119.

Gonnet, G.H., M.A. Cohen, and S.A. Benner, 1992, "Exhaustive matching of the entire protein sequence database," *Science* **256**, 1443-1445.

Greer, J., 1990, "Comparative modeling methods: Application to the family of the mammalian serine proteases," *Proteins Struct. Funct. Genet.* **7**, 317-334.

Gregoret, L.M., and F.E. Cohen, 1990, "Novel method for the rapid evaluation of packing in protein structures," *Journal of Molecular Biology* **211**, 959-974.

Hagler, A.T., and B. Honig, 1978, "On the formation of protein tertiary structure on a computer," *Proceedings of the National Academy of Sciences USA* **75**, 554-558.

Holley, L.H., and M. Karplus, 1989, "Protein secondary structure prediction with a neural network," *Proceedings of the National Academy of Sciences USA* **86**, 152-156.

Howard, A.E., and P.A. Kollman, 1988, "An analysis of current methodologies for conformational searching of complex molecules," *J. Med. Chem.* **31**, 1675-1679.

Jones, T.A., and S. Thirup, 1986, "Using known substructures in protein model building and crystallography," *EMBO J.* **5**, 819-822.

Kabsch, W., and C. Sander, 1984, "On the use of sequence homologies to predict protein structure: Identical pentapeptides can have completely different conformations," *Proceedings of the National Academy of Sciences USA* **81**, 1075-1078.

Karlin, S., and S.F. Altschul, 1990, "Methods for assessing the statistical significance of molecular sequence features by using general scoring schemes," *Proceedings of the National Academy of Sciences USA* **87**, 2264-2268.

Kauzmann, W., 1959, "Some factors in the interpretation of protein denaturation," *Adv. Protein Chem.* **14**, 1-63.

Kendrew, J.C., 1963, "Myoglobin and the structure of proteins," *Science* **139**, 1259-1266.

King, R.D., and M.J.E. Sternberg, 1990, "Machine learning approach for the prediction of protein secondary structure," *Journal of Molecular Biology* **216**, 441-457.

Kneller, D.G., F.E. Cohen, and R. Langridge, 1990, "Improvements in protein secondary structure prediction by an enhanced neural network," *Journal of Molecular Biology* **214,** 171-182.

Konigsberg, W.H., and H.M. Steinman, 1977, "Strategy and methods of sequence analysis," pp. 1-178 in *The Proteins,* Vol. 3, 3rd ed., H. Neurath and R.L. Hill (eds.), New York: Academic Press.

Kumamoto, C.A., 1991, "Molecular chaperones and protein translocation across the *Escherichia coli* inner membrane," *Molecular Microbiology* **5**, 19-22.

Kuwajima, K., 1989, "The molten globule state as a clue for understanding the folding and cooperativity of globular-protein structure," *Proteins Struct. Funct. Genet.* **6**, 87-103.

Lee, B., and F.M. Richards, 1971, "The interpretation of protein structures: Estimation of solvent accessibility," *Journal of Molecular Biology* **55**, 379-400.

Lesser, G.J., and G.D. Rose, 1990, "Hydrophobicity of amino acid subgroups in proteins," *Proteins Struct. Funct. Genet.* **8**, 6-13.

Levitt, M., 1976, "A simplified representation of protein structures and implications for protein folding," *Journal of Molecular Biology* **104**, 59-107.

Levitt, M., and C. Chothia, 1976, "Structural patterns in globular proteins," *Nature* **261**, 552-558.

Lifson, S., and A. Warshel, 1969, "Consistent force field for calculations of conformations, vibrational spectra, and enthalpies of cycloalkane and *n*-alkane molecules," *J. Chem. Phys.* **49**, 5116-5129.

Lim, V.I., 1974, "Structural principles of the globular organization of protein chains: A stereochemical theory of globular protein secondary structure," *Journal of Molecular Biology* **88**, 857-894.

Linderstrom-Lang, K.V., and J.A. Schellman, 1959, "Protein structure and enzyme activity," pp. 443-510 in *The Enzymes,* Vol. 1, P.D. Boyer (ed.), New York: Academic Press.

Maxam, A., and W. Gilbert, 1980, "Nucleic acids Part I," pp. 499-560 in *Methods in Enzymology*, Vol. 65, L. Grossman and K. Moldave (eds.), New York: Academic Press.

Murzin, A.G., and A.V. Finkelstein, 1988, "General architecture of the alpha-helical globule," *Journal of Molecular Biology* **204**, 749-769.

Novotny, J., A.A. Rashin, and R.E. Bruccoleri, 1988, "Criteria that discriminate between native proteins and incorrectly folded models," *Proteins Struct. Funct. Genet.* **4**, 19-30.

Overington, J., M.S. Johnson, A. Sali, and T. L. Blundell, 1990, "Tertiary structural constraints on protein evolutionary diversity: Templates, key residues and structure prediction," *Proceedings of the Royal Society of London, Series B: Biological Sciences* **241**, 132-145.

Padmanabhan, S., S. Marqusee, T. Ridgeway, T.M. Laue, and R.L. Baldwin, 1990, "Relative helix-forming tendencies of nonpolar amino acids," *Nature* **344**, 268-270.

Pauling, L., 1967, *The Chemical Bond: A Brief Introduction to Modern Structural Chemistry,* Ithaca, N.Y.: Cornell University Press.

Pauling, L., R.B. Corey, and H.R. Branson, 1951, "The structure of proteins: Two hydrogen-bonded helical configurations of the polypeptide chain," *Proceedings of the National Academy of Sciences USA* **37**, 205-211.

Pearl, L.H., and W.R. Taylor, 1987, "A structural model for the retroviral proteases," *Nature* **329**, 351-354.

Ponder, J.W., and F.M. Richards, 1987, "Tertiary templates for proteins: Use of packing criteria in the enumeration of allowed sequences for different structural classes," *Journal of Molecular Biology* **193**, 775-791.

Presnell, S.R., B.I. Cohen, and F.E. Cohen, 1992, "A segment based approach to protein structure prediction," *Biochemistry* **31**, 983-993.

Presnell, S.R., and F.E. Cohen, 1989, "Topological distribution of four-alpha-helix bundles," *Proceedings of the National Academy of Sciences USA* **86**, 6592-6596.

Ptitsyn, O.B., and A.A. Rashin, 1975, "A model of myoglobin self-organization," *Biophys. Chem.* **3**, 1-20.

Qian, N., and T.J. Sejnowski, 1988, "Predicting the secondary structure of globular proteins using neural network models," *Journal of Molecular Biology* **202**, 865-884.

Richards, F.M., 1977, "Areas, volumes, packing, and protein structure," *Annu. Rev. Biophys. Bioeng.* **6**, 151-176.

Richmond, T.J., 1984, "Solvent accessible surface area and excluded volume in proteins," *Journal of Molecular Biology* **176**, 63-89.

Richmond, T.J., and F.M. Richards, 1978, "Packing of α-helices: Geometrical constraints and contact areas," *Journal of Molecular Biology* **119**, 537-555.

Rumelhart, D.E., G.E. Hinton, and R.J. Williams, 1986, *Parallel Distributed Processing: Explorations in the Microstructure of Cognition,* Vol. 1, Cambridge, Mass.: MIT Press, pp. 318-362.

Sippl, M.J., 1990, "Calculation of conformational ensembles from potentials of mean force: An approach to the knowledge-based prediction of local structures in globular proteins," *Journal of Molecular Biology* **213**, 859-883.

Smith, R.F., and T.F. Smith, 1990, "Automation generation of primary sequence patterns from sets of related protein sequences," *Proceedings of the National Academy of Sciences USA* **87**, 118-122.

Smith, T.F., and M.S. Waterman, 1981, "Comparison of biosequences," *Adv. Appl. Math.* **2**, 482.

Sondek, J., and D. Shortle, 1990, "Accommodation of single amino acid insertions by the native state staphylococcal nuclease," *Proteins Struct. Funct. Genet.* **7**, 299-305.

Troyer, J.M., and F.E. Cohen, 1991, "Simplified models for understanding and predicting protein structure," pp. 57-80 in *Reviews in Computational Chemistry,* K.B. Lipkowitz and D.B. Boyd (eds.), New York: VCH Publishers, Inc.

Wendoloski, J.J., and F.R. Salemme, 1992, "PROBIT—A statistical approach to modeling proteins from partial coordinate data using substructure libraries," *J. Molec. Graphics* **10**, 124-126.

Wierenga, R.K., and W.G. Hol, 1993, "Predicted nucleotide binding properties of p21 protein and its cancer associated variant," *Nature* **302**, 842-844.

Wilson, C., and S. Doniach, 1989, "A computer model to dynamically simulate protein folding: Studies with Crambin," *Proteins Struct. Funct. Genet.* **6**, 193-209.

Wilson, C., L.M. Gregoret, and D.A. Agard, 1993, "Modeling side-chain conformation for homologous proteins using an energy-based rotamer search," *Journal of Molecular Biology* **229**, 996-1006.

Wuthrich, K., 1986, *NMR of Proteins and Nucleic Acids,* New York: John Wiley & Sons.

Appendix

Chapter Authors

Craig J. Benham
Professor and Chair, Biomathematical Sciences Department
Mount Sinai School of Medicine
New York, New York

Dr. Benham was trained as a pure mathematician, his thesis being in the field of complex manifold theory. He received his Ph.D. degree in mathematics from Princeton University in 1972. During his first academic appointment at Notre Dame University, he became interested in problems involving biomolecular structure. In order to educate himself in this new area, he took a postdoctoral position in 1976 with Max Delbrueck in the Biology Division at California Institute of Technology. At that time he began his work on superhelical DNA.

Dr. Benham initiated the theoretical analysis of superhelical DNA conformational equilibria. Over the years he has developed both elasto-mechanical and statistical approaches to the analysis of the large-scale structure of supercoiled DNAs. He also has developed theoretical methods to predict changes in the secondary structure of DNA in response to imposed stresses, the subject of Chapter 7 in this volume.

Fred E. Cohen
Professor of Pharmaceutical Chemistry, Medicine, Pharmacology, Biochemistry, and Biophysics
University of California
San Francisco, California

Dr. Cohen's research on computational approaches to the protein folding problem began as an undergraduate at Yale University and was the subject of his doctoral work at Oxford University, where he studied as a Rhodes Scholar. Upon completing graduate school, he came to the University of California, San Francisco (UCSF) as a postdoctoral fellow and simultaneously began medical school at Stanford University. He subsequently completed a medical residency and subspecialty training at UCSF before joining the full-time research faculty in 1988.

Dr. Cohen's research interests center around computational models for protein folding, structure, and function. Applications of this work have led to advances in drug design and discovery with a particular emphasis on parasitic disease. More recent work has attempted to understand the structural underpinnings of the neurodegenerative diseases caused by prions. Dr. Cohen has received a Searle Scholars Award, the Richard E. Weitzmann Young Investigator Award from the Endocrine Society, and the Young Investigator Award of the Western Society of Clinical Investigation. Dr. Cohen serves on the Molecular and Cellular Biophysics Study Section of the National Institutes of Health, as an editor of the *Journal of Molecular Biology*, and as a member of the editorial boards of *Protein Engineering, Perspectives in Drug Discovery and Design, Computational Biology,* and *Molecular Medicine.*

Eric S. Lander
Director, Whitehead Institute for Biomedical Research and
Professor, Massachusetts Institute of Technology
Cambridge, Massachusetts

Dr. Lander is a member of the Whitehead Institute for Biomedical Research, director of the Center for Genome Research, and professor of biology at the Massachusetts Institute of Technology. His background includes both pure mathematics and laboratory molecular genetics. He also taught managerial economics at the Harvard Business School from 1981 to 1989. Early in this period, he became interested in biology and acquired laboratory training in fruit fly, nematode, and human genetics.

Dr. Lander's theoretical work includes the development of mathematical methods for the genetic dissection of complex inherited traits; algorithms for genetic mapping; analytical approaches to physical map construction; and population genetic methods for finding human disease genes. His laboratory work includes the construction of genetic linkage maps of the mouse and rat genomes; construction of physical maps of the human genome; genetic dissection of traits, including colon cancer susceptibility in the mouse and diabetes susceptibility in the rat;

and cloning of human disease genes. He receive a MacArthur Fellowship in 1987 for his work at the interface of molecular biology and mathematics. He received a Ph.D. in pure mathematics from Oxford University in 1981, where he studied algebraic combinatorics.

Eugene W. Myers
Professor, Department of Computer Science
University of Arizona
Tucson, Arizona

Dr. Myers specializes in algorithm designs, an area of computer science focusing on the creation of computer methods to solve problems efficiently and accurately. He first entered the area of computational molecular biology in 1986 when he began to focus on algorithms for searching biosequence databases, discovering DNA sequence patterns, comparing sequences, sequencing DNA, and displaying molecular images. He is an editor for *Computer Applications in the Biosciences* and the *Journal of Computational Biology*. His algorithm designs are at the heart of the software tools BLAST, Inheret, ANREP, and MacMolecule.

Dr. Myers received a bachelors degree in mathematics from the California Institute of Technology in 1975 and a Ph.D. in computer science from the University of Colorado in 1981. Immediately thereafter he joined the faculty of the Department of Computer Science at the University of Arizona, where he became a full professor in 1991 and where he works today.

De Witt Sumners
Distinguished Research Professor
Florida State University
Tallahassee, Florida

Dr. Sumner's professional interests are knot theory and scientific applications of topology, and his research activity includes knotting in random chains and topological models in molecular biology. Dr. Sumners initiated the development of the tangle model to analyze the binding and mechanism of enzymes which alter the goemetry and topology of DNA. He has been a visiting professor at Kwansei Gakuin University in Japan and at the University of Geneva in Switzerland. He is a member of the Program in Mathematics and Molecular Biology at the University of California at Berkeley and has been a member of the Mathematical Sciences Research Institute, Berkeley and the Institute for Advanced Study, Princeton. Dr. Sumners received the B.Sc. degree in physics from Louisiana State University, Baton Rouge, in 1963 and the Ph.D. in mathematics (specializing in topology) from the University of Cambridge in 1967, where he was a Marshall Scholar. Dr. Sumners is a member of the editorial boards of the *Journal of Knot Theory and Its Ramifications, Nonlinear World*, and the *Journal of Computational Biology*.

Simon Tavaré
Professor of Mathematics and Biological Sciences
University of Southern California
Los Angeles, California

Dr. Tavaré's scientific interests are in the application of probability and statistics to problems in population genetics, human genetics, and molecular evolution. He has held positions at the University of Utah and Colorado State University. Dr. Tavaré is a Fellow of the Institute of Mathematical Statistics. He received a Ph.D. in probability and statistics in 1979 from the University of Sheffield in England.

Michael S. Waterman
Professor of Mathematics and Biological Sciences
University of Southern California
Los Angeles, California

Dr. Waterman's main scientific interests are in the application of mathematics, statistics, and computer science to molecular sequence data. He holds a USC Associates Endowed Chair and is a Fellow of the Institute of Mathematical Statistics and a Fellow of the American Association for the Advancement of Science. Dr. Waterman has held positions at Idaho State University and Los Alamos National Laboratory. He received a Ph.D. in probability and statistics from Michigan State University in 1969.

James H. White
Professor, Department of Mathematics
University of California
Los Angeles, California

Professor White's work involves the study of the geometric properties of curves and surfaces in three-space. It was in his thesis that he gave the first complete mathematical proof (outlined in the first part of Chapter 6 in this volume) that the linking number of the backbone strands of a closed circular DNA is equal to the twist of one of the backbone strands about the axis plus the writhing number of the axis. Because of this work, he was contacted in 1977 by Francis Crick at the Salk Institute to explain how the supercoiling of the axis of a DNA affected its topology. The answer to Crick's question is included in the second part of Chapter 6, having evolved over the years not only to applications to DNA but also to DNA-protein interactions.

Professor White has been involved with mathematical applications to DNA for almost 18 years, assisting molecular biologists in many laboratories throughout the United States and Europe, including William Bauer at SUNY, Stony Brook and Nicholas Cozzarelli at the University of California, Berkeley. These collaborations have led to the development of the surface linking theory outlined in Chapter 6 and the applications of knot polynomial theory to recombination.

Professor White is a founding member of the Program in Mathematics and Molecular Biology, a National Science Foundation project that has been established to further the advancement of the mathematical sciences in the field of molecular biology. As a member of this group, he has written many articles and has organized several conferences on the applications of geometry and topology to DNA and protein work. He is at the forefront of those encouraging young researchers to work in this new interdisciplinary field. He received a Ph.D. from the University of Minnesota in 1968 in the field of differential geometry.

Index